VALENTIN MALINOV

MYTHS LIES ILLUSIONS AND THE WAY OUT

FACTS ABOUT THE GREATEST SCIENTIFIC DECEPTIONS OF OUR TIME

MYTHS LIES ILLUSIONS AND THE WAY OUT

First published in Australia by Valentin Malinov 2018 (1st Edition)

2nd edition published in Australia by Valentin Malinov 2020

Copyright © Valentin Malinov 2020
All Rights Reserved

ISBN: 978-0-6480127-2-6 (pbk)
ISBN: 978-0-6480127-3-3 (ebk)
Details available from The National Library of Australia

Artwork and photography by Valentin Malinov © 2018

Typesetting and design by Publicious Book Publishing
Published in collaboration with Publicious Book Publishing
www.publicious.com.au

**This book is dedicated to the genius of Nikola Tesla-
the man, whose contribution to humanity
is unsurpassed to present day**

In science, you cannot explain something you don't know
with something you don't understand.
(V. Malinov)

CONTENTS:

INTRODUCTION

There are a few fundamental questions in our lives: Who are we, and where do we come from? What is the purpose of our life? Where does the universe come from? Does God exist? Is there intelligent life in the universe? Are we alone? What is our future?

I was curious to learn the answers to these questions, and the more I have learned, the more confused and disoriented I became. There are countless different and confusing explanations and theories for all aspects of our lives, but I didn't find any satisfactory, logical, and credible explanations for any of these simple but vital questions.

I would like to tell you of my struggle to find truthful answers to these questions. I will start my explanation with a strange question: what is tyranny? In everyone's mind, tyranny is associated with a cruel dictator who terrorizes, and mercilessly kills people. This is correct, but this is the extremity of tyranny. There is another form of tyranny that is stealthier, hidden but not less cruel than the first category. This form of tyranny, whose real extent is difficult to be determined, is what I call 'social' tyranny, where a selected minority group succeeds in establishing a corrupt system with the aim to exploit the rest for their own benefit. The social form of tyranny mainly uses tactics of deception to promote a corrupt system where the rule of law applies to everyone else, but 'they' are above the law, or creating laws that benefit only their group. They have succeeded in labeling this form of social tyranny as "the best democratic system", which is possible to have. Tyrannies have long roots in human history and were able to produce and establish very sophisticated systems for deception and suppression of any idea that would expose the roots of the corrupted systems of social tyranny.

The reason for this book is to expose the system of deception imposed on us and to reveal the true extend of the well-orchestrated system of false knowledge, wrong information, and suppression of science. It is necessary to explain why there are so many opposing theories, facts, and stories in all aspects of life, science, history, philosophy, and religions because the distortional perception of the world and reality is a mortal and imminent danger for every living organism, any individual, any institution, and any civilized society!

The necessity to write this book gradually gained strength in parallel with my knowledge and understanding of what's really is going on in our society and the world. I was shocked when I learned that scientists in the early 20th century found undeniable evidence for "God" or (Creator) existence with the new discoveries in quantum mechanics, physics and astronomy. I was so innocent and naïve that I didn't understand why such vital discoveries were immediately suppressed and hidden from the public. To have answers to these questions, I seriously studied the social, scientific, political, and economic structures of the world. The real picture slowly started emerging from the fog of chaos, misinformation, and confusion.

From history, we know that dogma and tyranny are closely linked. The old and proven strategy of conquering a nation and enslaving its people is by eliminating that nation's leading intellectuals, its ethnic unity, and its cultural, moral, and religious identity. Only when this is done, and the nation loses its intellectuals, social bonds, its unique culture, and identity, only then can it be ruled by force, lies, and deception!

If you start analyzing the facts, it won't be very difficult to understand that at present, this is exactly what's happening.

The presence of well-organized forces behind the scenes that possess enormous wealth and have for a long time been in control of the economic, political, moral and social systems of the western world, is very obvious. To maintain their dominance, the members of the ruling society adopt a strategy for social, cultural, ethnic, religious and international disunity and confrontation, because well-organized people and countries are a threat to their rule. They adopt policies for total misinformation and confusion over all aspects of life. They manage to pit the different social and ethnic groups against each other, such as science against religion. They orchestrate religious tensions and social, cultural, moral, and ethnic intolerance. They succeed in dividing nations, denying the rights of self-determination of millions of people and ethnic groups. They have destroyed families, the foundation where young people receive knowledge of right and wrong. The social media in different countries is bombarding people with opposing sides of the stories and make them hate and blame each other. In this situation, a united scientific community has the potential to reveal the truth and to unite the intelligent people of the world. That's why the first thing rulers do is to suppress knowledge and science.

Science has succumbed to enormous pressure and enters some kind of new 'Dark Age.' In order to have funds and jobs, the scientists must accept and obey the rules and agenda imposed from above. It is a pity to see the result of such coordinated attacks and interference in science, where the real facts and results are hidden, twisted and suppressed. A green light is given to cleverly formulated nonsense, which creates confusion in all aspects of knowledge and life. The basic idea of religious beliefs is – that people should be honest, intolerant to corruption and to live in peace, moral and harmony – is very uncomfortable for the ruling super-rich elite, because they gain their wealth and power by using unfair laws and tactics, deception, wars, exploitation of natural and mineral recourses, manipulation of the market, creating inflations, recessions and wars. All this activity is well hidden behind information chaos and confusion, with its social, ethnic, religious, and international tension and confrontation. That's why all the scientific facts for human history, the knowledge of the world, for the existence of a creator of the universe, or (God), the real principles of democracy and moral and ethical standard are unacceptable to them and is heavily suppressed.

How do they manage to do this? Simple! They are using public money to fund the deception – (our money). Through government funding for science, the ruling elite direct and promote only the programs and research they approve of and publish only what is serving their agenda. It is a well-known fact that only selected scientists to have the privilege of making public announcements; the media is silent to the voices of scientists who express concern and disagree with the official line of the story, findings or theory. To uncover the true extent of the suppression of science, later in this book I will go into details and discredit many bizarre and absurd officially promoted mainstream pseudo-scientific findings and theories which oppose the laws of physics, the laws of nature and common sense. I am not trying to promote my views, beliefs, or agenda, neither to promote nor denounce belief in God. I would like to present the facts that are hidden from the public – all the facts without preferences. I will leave the readers to draw their own conclusions after reading my findings of the false theories and misconceptions in current science, where every explanation is full of conflicting properties, incorrect facts and nonsense that opposes any logic, observational data or natural and physical laws. They usually cover up the nonsense with complicated mathematical calculations, but this nonsense will be exposed in easily to understood, everyday language, and will be in a scientific way that will be in accordance with the laws of physics, observational and experimental data, logic and common sense.

THE SUPPRESSION OF SCIENCE

Ancient civilizations, including the Greek civilization, had remarkable scientific knowledge in mathematics, physics, philosophy, and astronomy. The Greek philosophers have described the atomic structure of matter with remarkable accuracy compatible with the present understanding. They knew that not the earth, but the sun is the center of the solar system. The Greek and Egyptians scientists were able to calculate the size of the earth, the moon and the sun and the distances between them, including all visible planets in the solar system with remarkable precision. Two bronze celestial clocks have been found, containing a number of toothed gear wheels that simultaneously show the movement and position of all visible planets with accuracy up to seven decimal points.

The astronomical clock of the ancient Greeks

The strategy of tyranny is to destroy the knowledge and the leading intellectuals in order to be able to deceive and rule. This has started with Athenians, who was sentenced to dead Socrates, who dare to teach that morals and ethics are the principle of life and Universe. There is not an accident that the first action of the conquering Roman Empire was to kill scientists like Archimedes and to burn the library in Alexandria. They already knew that knowledge is dangerous for tyrannies. The systematic destruction of knowledge continued with the establishment of Christianity, where scientific books were methodically burned in the public square, and the scientists branded as heretics, and devil servants. After a few centuries, people ended up with only one available book– the Bible.

I don't want to offend the feelings of people who believe in and respect the Bible. Currently, there is absolute silence when and how the New Testament was produced. I read Christian history forty years ago and am puzzled why these historical facts are hidden. The New Testament was produced under the guidelines and supervision of the Roman Emperor Constantine in the 4th century. The Emperor arranged a meeting of selected by him religious leaders in Nicaea, with the aim to unify Christian beliefs and use it for his political reasons. As a result, he produced the New Testament. That's why, to be accurate, any Bible translation must be from the Greek, not from Latin, because the Greek language was the official one in Constantinople. The tragedy is that, even in the most precious book of Christianity, the ruling authorities managed to insert their agenda. Under the supervision of the emperor, who proclaimed himself to be God's representative on the Earth, in the New Testament, there were included only the writings of the apostles with which the Emperor agreed. The opponents were exiled, and their work and chapters burned. For the head of the government that killed Jesus Christ and persecuted his followers for three centuries, to proclaim himself God's representative on Earth and to direct and arrange the assembling of the Christian holy book is supreme arrogance.

Emperor Constantine and his icon as God's representative on earth

Under his supervision, the Christian church continues to maintaining human slavery, claiming that Jesus never criticized the slavery system!

The English translation of the New Testament is made by William Tyndale from Greek in 1500. For the 'crime' that he makes the Bible available to the people, he was executed in a most barbaric way because the people won't need anymore the priests' interpretation of the meaning of the Bible. Anyway, the damage has been done, and in1604 -1611 under the supervision of King James the translation has been edited and adapted to the purpose of the authority. The story is repeated again! - The King inserted his political agenda in the new translation and reserve for himself the position - head of the church, maintaining the claim that the authority to rule the people is given and blessed by God! Under the authority of the 'God's representative, the new wave of transatlantic slave trade toward America was established, tolerated, and legislated!

We will never know how many good biblical chapters have been erased. The result of this intervention is well known – under cover of modesty, ethics, and love is carefully inserted an agenda of obedience to authority, tolerance to tyranny, cruelty, and injustice, and where people are taught to be patient and obedient and, if they suffer, they must wait patiently to die and be rewarded in heaven. Personally, I don't believe that tolerance of tyranny, cruelty, and injustice could be really God's instructions. I believe it to be exactly the opposite! Where the ruling authority has inserted the idea that instead of resist tyranny and corruption, the people must be obedient and wait for God to intervene and save them!

The ruling elite started using the church as a tool of suppression over a period of more than a thousand years. The dogma of the Catholic Church was enforced as a scientific explanation for everything. We call this period the Dark Ages because in this period the scientific books were banned and burned, scientific discussions were branded as the devil's ideas, some scientists were tortured and burned at stake, and others were forced to denounce their ideas publicly.

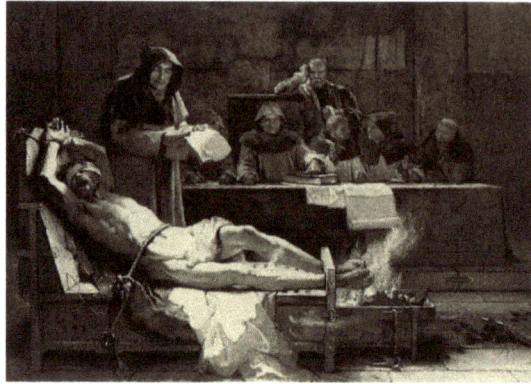

How the Inquisition dealt with science

During this period, one of the 'scientific' discussions was how many devils could stand on top of a pin. And they proved 'beyond doubt' that forty devils can stand on top of a pin. You can only reach such nonsense when you start ignoring the truth, the real facts, and thereby lose a sense of reality. My concern is, are we again creating a new Dark Age with the current suppression of science, truth, and knowledge? We are witnessing that any source exposing the false official version is branded like 'conspiracy theory' or 'fake news' – same as the 'heresy' of the past.

The foundation of modern physics was established early in the20th century. The concept of quantum mechanics was written at the Solvay Conference in 1927 in Copenhagen, coincidently again under the supervision of the ruling elite. This conference was funded by a wealthy private sponsor and, as a result of this, it produced the most arrogant and shameless document in the history of science, - the 'Copenhagen Interpretation,' this is doctrine for the future development of quantum mechanics and for our understanding of the world. The document stated that: - quantum mechanics has a complete explanation of the world, and future knowledge is not possible. Further search for the fundamental properties of the world is not necessary!

I cannot see any difference between the Dark Ages' religious dogma and this shameless document, because this doctrine was produced when the science had little or no any understanding of the fundamental property of our world. And as a result of this document, the funds for future research were allocated carefully in a direction where scientists will never be able to find the fundamental properties and laws of nature. In physics a slogan was inserted– 'shut up and calculate.' Calculate what? Lack of knowledge with invented numbers? Albert Einstein never agreed with this document and famously said; 'I don't have to look up to know that the moon is there' (because Niels Bohr claimed that nothing is real). And again, is it a coincidence that this shameless document came as a result of a privately funded institute, with a privately funded director, - (Niels Bohr), and at a privately funded conference in a private hotel? You can draw your own conclusions for the ethical standards of the two leading figures standing on the opposing side of this story. Niels Bohr, - who volunteered to help in the development of the atomic bomb, and Einstein who declined to participate in producing this terrible weapon and to participate in politics. - (He also declined the offer to be prime minister of Israel.) He even wrote a letter to Roosevelt to convince him not to develop the atomic bomb.

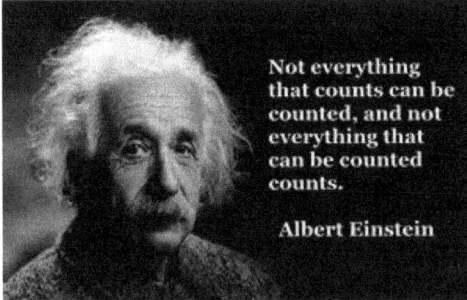

A. Einstein   Niels Bohr

Here is the story of how the greatest man in modern physics - Albert Einstein, was blackmailed and forced to abandon his research and to support the most bizarre and regressive theory for our understanding of the world. In post-World War One Germany, anti-Semitism was gaining pace. Hitler's followers started attacking and banning anything that had Jewish associations. In this climate, Einstein, with good reason, feared for his life. Help came from an unexpected source – the Catholic Church in the form of the Belgian priest, Georges Lemaître. We can just guess what drama arose there. Georges Lemaître had proposed his theory of the Big Bang, but the scientific community hasn't had been too enthusiastic in accepting it because the theory has no real scientific ground. The only evidence in support of this theory is Hubble's claim of the light red-shift from the galaxies. This red-shift, according to Hubble, is supposed to indicate that the galaxies are moving away from us. The leading scientists at that time weren't that naïve in believing the claim, and even Hubble himself, sometime later when he became familiar with the theory of relativity and gravitational light red-shift denounced his claim, but his voice has been muted.

In1933 Einstein and Lemaître went to the United States together. There, Lemaître gave a public lecture to promote his theory. After the lecture, there was silence, which was suddenly interrupted by the applause of Einstein, who rose to his feet and proclaimed that he had never heard a more beautiful creation story–'story' is not facts. But the respect due to Einstein was crucial for the acceptance of this theory. We have to take into account the circumstances surrounding this event. Einstein was in a situation where his life was in danger if he could not immigrate to the US. In this situation, the support of the Catholic Church was crucial. Einstein was allowed to stay in the US and was given a position in Princeton, New Jersey. Einstein definitely knew that this hypothetical recession of the galaxies was not correct. It is more than obvious that he'd been forced to accept such a fundamentally incorrect assumption. He told Lemaître, 'Your mathematics is OK, but your physics is awful.' What kind of deal he had been forced to accept we cannot know, but the fact that he stopped his research and never published any continuation of the Theory of Relativity or any other fundamental aspect can tell you the story! At least he had been free to oppose the nonsense of the Copenhagen Interpretation.

Humanity can be proud of its action: - we managed to humiliate and succeed in stopping the work of one of the greatest minds of our time and muted him.

Albert Einstein and Georges Lemaître

Can it be an accident that from the same place where Einstein was working, a young American physicist, David Bohm, dared to challenge the view of Niels Bohr and produced good evidence that physics can describe reality? For this, he was fired from Princeton University and not given any academic jobs in the United States, despite that his work leads to the famous Bell theorem. - This is the reason for my concern because we are doing exactly the same bad things as in the Dark Ages: we are giving the green light to any nonsense, which comes from authority-selected people who promote confusion, ignore laws, disobey common sense, and suppress evidence and reasonable ideas. Their opponents are banned from publications and public speeches, denied access to public facilities for research; they are disqualified and are not allowed to hold public offices or teach students. That's what also happened to the top astronomer, Alton Arp Halton, who dared to say that the observation data for quasars challenges the correctness of the Hubble constant. That's what also happens to thousands of honest scientists brave enough to have a different opinion.

I would like to explain briefly what the Hubble constant is because this is the basic equation behind our present picture of our world. The Hubble constant measures the speed of stars by measuring the stretching of light waves as a result of the movement of the star. The problem is that this equation does no include the other two factors that produce the same effect, the gravity field,

11

which Einstein very clearly described in the Theory of Relativity and the electromagnetic field of the universe. I will insert here a section from Wikipedia about the history for the 'correctness' of this constant:

Edwin Hubble

"In 1927, two years before Hubble published his article, the Belgian priest and astronomer Georges Lemaître was the first to publish research deriving what is now known as Hubble's Law. Unfortunately, for an unknown reason, "all discussions of radial velocities and distances (and the very first empirical determination of 'H') were omitted." There are speculations that these omissions were deliberate.

According to Canadian astronomer Sidney van den Bergh, "The 1927 discovery of the expansion of the universe by Lemaître was published in French in a low-impact journal. In the 1931 high-impact English translation of this article, a critical equation was changed by omitting reference to what is now known as the Hubble constant. That the section of the text of this paper dealing with the expansion of the universe was also deleted from that English translation suggests a deliberate omission by an unknown translator." - (No comment!)

For the 'accuracy and reliability' of Hubble constant, you can make your own conclusion, when you know the real facts behind. When the first value of the constant was applied in 1931, the age of the universe come to be 1.9 billion years! Since then, the constant has been "adjusted' about 27 times, to be able to have acceptable value. In 1956 Humason and Alan Sandage succeed to adjust it to 5.6 billion years age of the universe, and after a few further adjustments, we have the present value of the Hubble constant, which has been reduced numerous times, to fit the number, which they needed!

In *New Scientist* magazine, 22–28 May 2004, a letter from 34 scientists was published, saying that any criticism of the Big Bang theory is effectively blocked. This letter also was covered in silence.

A well-known fact is that inventions for cheap or free energy are heavily suppressed, research funds denied, and many of the inventors even disappear. The Japanese satellite Hitomi, worth $260 million with a capability to see further than the officially promoted visible age of our universe, was successfully launched, but shortly after, in April 2016, for some 'mysterious' reason, contact with the satellite was lost. We lost the ability to see is the promoted end of the universe is real or if there is more universe behind to be seeing. This situation is very like the Dark Ages described above - an orchestrated suppression of science and knowledge at any cost. This is scientific vandalism!

On the first page of the book, *Quantum Enigma*, by Bruce Rosenblum and Fred Kuttner, the authors shared the experience of their interview with Einstein in the 1950s:

'Einstein soon asked about our quantum mechanics course. He approved of our professor's choice of David Bohm's book as the text, and he asked how we liked Bohm's treatment of the strangeness quantum theory implied. We couldn't answer. We'd been told to skip that part of the book and concentrate on the section titled "The Mathematical Formulation of the Theory." Einstein persisted in exploring our thoughts about what the theory really meant. But issues that concerned him were unfamiliar to us. Our quantum physics courses focused on the use of the theory, not its meaning. Our response to his probing disappointed Einstein and that part of our conversation ended.'

This passage doesn't need explanation; it explains well how the nonsense has been established and maintain.

Mathematicians have been promoted to leading positions, despite do not have a good understanding of physics and are likely to disregard the laws of physics. For them, it is very easy and logical to put a minus sign in their equations, and use assumed and selected values and properties. We know that in mathematics if you put in a slightly incorrect number, the mistake gradually gets bigger and bigger with the progress of calculations and you get bizarre results.

I will give some examples. With mathematics, you can easily calculate properties below −273 degrees, but in physics, you cannot go below absolute zero, because this is the lowest level of energy, but for mathematicians that means nothing. The mathematicians give us 'dark energy' which is negative gravity. Every physicist knows that you cannot put in a closed physical system (the universe) values with opposing properties like (+) and (−) because they will cancel each other. But mathematicians don't care that that violates to an extreme extent the laws of physics. It is a similar situation with the claim that the universe comes from nothing because in mathematics you can divide '0' to any two equal sums with positive and negative signs, but in reality, from nothing coming out only nothing! The universe is a closed physical system, and antimatter has nowhere to go! Anyway, even this was no obstacle for mathematicians, who claim that the universe comes from nothing for no apparent reason and that antimatter, which is supposed to be an equal amount to normal matter, for some mysterious reason has disappeared. Are they OK? - Great assumption with 50% of missing matter, and 100% of missing understanding of the laws of physics and reality that from nothing comes only nothing! The pseudo-scientists have established a constant practice of announcing anomalies, which do not fit the standard model and ask for additional funds to study those non−existing phenomena. Astrophysics has lost direction and any credibility. It turns itself into a public show of fantasy and starts producing unbelievable, stupid theories which have no scientific or experimental support: The Big Bang Theory, String Theory, 'the observer creates the reality', the holographic universe, multi-verse, the god particle, mass-less particles, dark matter, dark energy, the CMB from Big Bang, the universe from nothing. Well done, congratulations, you are geniuses!

The elite succeeds in making complete chaos of our understanding of the structure of the world, the origin of life, human origins, and history. They manage to turn nations against nations, religion against science, and science against religion. They have destroyed the basic principles of civilized societies – ethics, morals, and tolerance.

In the contra-argument of these facts, they immediately will start to question my qualifications, but they will not comment on the subjects. Why? I would like to say that qualifications are no guarantee for real intelligence, honesty, and knowledge. You can always learn more from one honest person than from a bunch of educated crooks. Not many people know the difference between educated and intelligent people. Actually, the level of education and knowledge is not a real indication for the person's intelligence. Most of the university students possessing more knowledge than Archimedes, or Aristotle, but that doesn't mean that the students are more intelligent! With their limited scientific data, those colossuses of the human intellect had a better understanding of the world than most of us! They show us that high ethical standard is a privilege only of intelligent minded people and have nothing to do with the educated crooks!

SCIENCE AGAINST RELIGION?

The ruling authority has succeeded in establishing the current view that science and religion are incompatible and in opposition to each other. Let's examine if this is true and correct, if there is a conflict and if conflict is where the actual conflict is.?

Conflicting situations will be if we have two opposing theories, two opposing facts or views. I will start with the fact that many leading scientists from the past and present believe in God. Nobody can say that people such as Newton, Faraday, Einstein, and Planck have less knowledge than us. The scientists from the past had no conflict in being scientists and believing in God. What has changed this situation? In the present time, many leading scientists are forced to hide their belief in God simply because they will lose their jobs. This fact is a pure act of tyranny and vandalism and has nothing to do with any scientific principles. In the present day, the top scientists are forced to state that in science, there is no room for religion. I am sorry, but such pre-determined statements are pure ignorance of the available facts and data. It is also a violation of the principle of science! - Those people don't deserve to call themselves scientists.

To consider a subject with a pre-determined position is not acceptable to science or scientific methods. This is a political approach, not a scientific one. Scientific methods are based on a consideration of facts, observations, experimental and statistical data, and logical analysis of all facts.

Max Planck is giving us a present understanding of the microscopic structure of matter, and Einstein is giving us the knowledge of the grand structure of the universe where all properties are related to each other.

The intelligence and understanding of the world of those two men are unquestionable, and we have to take notice of what they are trying to tell us:

'Both religion and science require belief in God. For believers, God is the beginning, and for physicists, He is at the end of all considerations. I regard consciousness as fundamental. I regard matter as derivative from consciousness. We cannot get behind consciousness. Everything that we talk about, everything that we regard as existing, postulates consciousness.'

(Max Planck)

'Everyone who is seriously engaged in the pursuit of science becomes convinced that a spirit is manifest in the laws of the universe – a spirit vastly superior to that of man, and one in the face of which we with our modest powers must feel humble.'

'I want to know how God created this world. I am not interested in this or that phenomenon, in the spectrum of this or that element. I want to know His thoughts; the rest are details.'(A. Einstein)

We have to learn from the sentences of those two great scientists who had the deepest understanding of the property and functions of our world.

The present science has found that the world is much more complicated and in the structure of nature are embedded nonphysical phenomenon, which are the driving forces behind the scene.

The science has better tools to describe the physical aspect of the world, but is hopeless to explain the nature of consciousness, the hidden law of physics, the invisible connection of every part of matter and universe, the phenomena conscious connection between all living organisms and our emotional and spiritual internal world!

The religions have a better way to explain the spiritual aspect of the world. Those two branches of human knowledge – science and religions have the same purpose - to explain the structure and purpose of our world. The aim of science and religion is the same, and they must accomplish each other. It is a pity to witness how the corrupted establishment has put wage between those two brunches of our knowledge.

The current scientific standard for findings to be recognized as correct requires certainty to five sigmas, or (five decimal points). What is the situation with the scientific evidence for spontaneous origin versus the intelligent design of our universe? The leading scientists, when considering the fine-tuning of the cosmological constants, forces, and properties of the universe, calculate that the chances of the universe being spontaneously formed versus intelligently designed are one against 10^{123}! - Do you realize the magnitude and certainty of this data? - Those finding of the universe to be intelligently designed has certainty of 99.99999... followed by 123 nines! By any scientific standard, this is certainty beyond any doubt. And when you have statistical evidence of such magnitude, to go public and make opposite pre-determined statements is absolutely unacceptable behavior for present scientists!

On the other hand, religions have arisen because there really is something mystical in nature, which many people are sensing but cannot explain. We - humans, are logical creatures, and we know that, for everything to exist, it must have purpose and reason. The world looks to be driven by invisible forces and logic. Religion is more philosophical than the precise explanation of nature and the world. The ruling elite is using religion for their purpose, but at the same time, they make everything possible to discredits it. This also is the case with the Theory of Evolution. This brilliant idea has masses of evidence that species are really evolving, but this theory was promoted like "The origin of spicy," and the Churches stubbornly rejected even the obvious facts of evolution. This argument has discredited both sides because both are incorrect in their fundamental assumptions. In reality, there is no conflict because evolution is correct, the species are evolving, but this cannot explain the origin of life. The church is correct that evolution is not the answer to the origin of life, but is incorrect about the evolving of species and the age of the earth. If both sides were more careful and modest in their assumptions, there would be no ground for argument. And this is the view of many scientists and many educated and religious people. Albert Einstein said, '**Science without religion is lame, religion without science is blind.**'

The unification point for science and religion is exactly that phenomenon in which the ruling authority hides at any cost - the true nature of the physical world. –

In the structure of the world and matter undoubtedly manifests the presence of intelligence, and the physical nature and structure of the world are neither far from religious nor scientific descriptions. There is no conflict at a fundamental level between science and religion because both of them are trying to unlock the secret of nature.

The structure of the universe contains three non-material elements – space, time, and information. The matter is the only material part of the system of nature. Science recognizes that matter is governed by the laws of physics, which manifest in universal information and consciousness. This description is not very far from the spiritual fundamentals of religion. The authorities succeed in denying all these non-material parts of reality and strip us of any chance for a reasonable understanding of the world. This is a pseudo-science, crippled, and adjusted to sick philosophy that has nothing to do with reality and real scientific knowledge.

It is a similar situation with religion. It has been divided into many, each opposing each other: Catholics, Orthodox, Protestants, Baptists, Jehovah's Witnesses, Buddhists, Hindus, and on and on. The institution of the church has departed from its main reason to serve the truth, its people and the idea of religion, and has allied with the authorities, no matter who they are and what their moral principles are. The church became an instrument of terror, with the Inquisition, burning scientific books, organizing military invasions like the Crusades, and even aligning with Hitler. (I am speaking of the institution of the church, not of ordinary people who believe and serve the good cause and ethical reason.) In a situation like this, people are very likely to get confused and turn their backs on religion and its fundamental ethical values and principles.

The institution of the church stubbornly rejects scientific discoveries and principles, despite science not opposing religious principles and proving beyond doubt that the basic principle of the religion – that in the fundamentals of our world lie intelligence and consciousness and they are embedded in our reality. Consciousness and intelligence are the software for the harmony and order of our universe. It is logical to ask, is this sign of the presence of God? Probably, but I will try to avoid inserting the term 'God' into scientific explanations.

I just have to say; the fact is that consciousness can only be the product of an intelligent mind! It will be good if we accumulate more knowledge with the aim to have a better understanding of the structure and nature of our world. But even with the facts and knowledge that we possess, we can have our own credible opinion about this subject. My point is that the church institution has departed from its purpose and values and become part of the orchestrated public confusion, is not defending basic moral, ethical and social values but vigorously opposing scientific facts and progress.

At the same time, science also has departed from its value and purpose. Mainstream scientists started serving politics and hiding the basic functions and principles of the physical world. As a consequence, we have ended with a situation where pseudo-science and pseudo-religion oppose each other and serve the orchestrated campaign to create chaos and confusion. We have to realize that science and religion have common fundamental grounds and purposes. - Both of them is explaining the world. The religion is concentrating on the spiritual aspect of the World; the science is obsessed with the physical aspect of the World. But the Universe is constructed on physical and non-physical elements, and science and religion have to unite their effort to present us with a complete picture of the world, not only spiritual, or just physical! Science and religion have the same purpose, and they definitely do not oppose each other!

FINDINGS IN QUANTUM MECHANICS IN THE EARLY 20TH CENTURY

You are probably wondering why I allocate so much space to physics. The answer is simple: because we live in a physical world and physics is the science which has the most credible explanation of our world. I will try to explain it in easy, every day, simple language.

The foundation of modern physics starts emerging in the second half of the 19th century, and by the beginning of the 20thcentury the principles of thermodynamics, electrodynamics, quantum mechanics and the Theory of Relativity had been established. This was a great time for the brightest minds of our time, which started to unlock, one by one, the secrets and structure of nature. The knowledge of the world started accumulating at an unbelievable pace. In contrast with the 'Middle Ages' picture of a simple world, new, a vast and complicated world started to emerge, filled with surprises, new laws, and properties, mysteries and puzzles, waiting to be explored. In the new discoveries of quantum mechanics, the strange, mystical, and inexplicable properties of the micro-world started emerging and led to a total rethinking of everything we know about the world. These strange and mystical properties of the micro-world are affected by our consciousness, and this can be explained only by the presence of a hidden order of physical laws, one behind the scenes. The special balance, logic, and ethics of these hidden physical laws lead to the conclusion that they can be only a product of super-intelligence. And the public knowledge of this emerging phenomenon is what the corrupt elite are afraid of.

The Solvay conference in 1927.
Einstein, Plank, and Schrödinger couldn't do anything to stop the crooks.

To hide the knowledge of the new discoveries, the elite immediately organized a scientific cop led by Bohr, Pauli, and Heisenberg. They succeed to insert a new philosophy in physics with the well-known 'Copenhagen interpretation' and the appearance of Big Bang Theory. In the new doctrine of physics was also inserted a new slogan – 'shut up and calculate.'

And this sentence became the slogan of the new era of scientific nonsense and degradation. Mathematicians were appointed to leading positions in physics, and they start "adjusting" the property of the universe to their mathematical equations.

The problem is there, where the mathematical formulas have no meaning and cannot explain the physical world. Physics has its own laws, which is not always logical and not obeying mathematics! In Physics, some particles can be 1 and 0 simultaneously, but in mathematics, every number have to define value. In mathematics, you can divide any number on equal positive and negative value

($0 = +10 - 10$) but in Physics you cannot do this!

In Physics is no such thing as negative energy, negative matter, negative time, negative space, negative force! Even antimatter has positive mass and energy! In mathematics, you can take any negative number, but in Physics you cannot do such things! In Physics, you cannot go lower than -273C. In mathematics - 1+1=0, but in physics most of the times -1+1=2!

(See the explanation below)

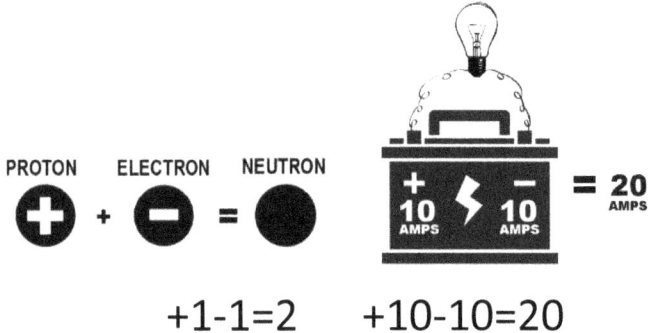

PROTON   ELECTRON   NEUTRON

+1-1=2   +10-10=20

The financial elite through various governments immediately withdrew all funding for continuing research for the fundamentals of quantum mechanics. The scientists were ordered to skip the research for the fundamental properties of matter and universe and to start smashing charged particles with the hope of finding the answer to all fundamental questions. This senseless direction leads to a dead-end, mess, and confusion – exactly, what the corrupted elite need.

As a consequence of this, quantum mechanics has had virtually no progress in the last hundred years. Even in present times, there is no research for finding the fundamental properties of the universe and matter – the origin of Information, the origin of laws of physics, the interaction of consciousness and matter, gravity, electromagnetism, nuclear forces, space, time, the structure and stability of the atom and so on. The scientists are only allowed to smash particles and to fantasize and produce bizarre theories. For this, they are well paid, well respected, and awarded with various scientific and Nobel Prizes. Leading scientist and Nobel Prize winner Richard Feynman has stated, '**I think I can safely say that nobody understands quantum mechanics. And if somebody claims that he understands quantum mechanics, I will call him a liar**'.

Unfortunately, this is the present state and tragedy in physics. You cannot calculate something you don't understand and is not possess physical property – as time, space, consciousness.

The honest, logical consideration of universal property, facts, data, and experiments has been abandoned! The scientists are allowed just to smash particles and use selected data and wire calculations to bombard us with bizarre impossible theories!

What actually were the important discoveries in the early 20th century?

To understand the importance of those findings, we have to start with an understanding of some of the fundamental elements, properties, and functions of the microscopic world. Classical physics deals mostly with bigger structures and describes the world with nearly perfect accuracy. There have been some minor puzzles and anomalies, but in general, they haven't been big problems. However, the new discoveries of the structure of atoms, the structure of matter and physical functions in the microscopic world bring new, revolutionary and better understandings of the world and answers to many puzzles of physics.

Albert Einstein proposed something unimaginable: that space, time, matter, and all forces are related and are dependent on each other, together they are forming one system where everything is made from energy. (Matter is the concentrated form of energy). In around 1923 the atomic structures were established and accepted. The German physicist Max Planck realized that the thermal energy of matter is absorbed and released in the form of photons. The photon actually is a package of energy with certain quantities or quanta. This discovery gives the name to the new branch of physics – 'quantum mechanics'.

The key experiment for the fundamental property of particles, their behavior, and the fundamentals of quantum mechanics is the double-slit experiment.

A beam of light passes through two parallel slits in a board. If the light is in the form of photons (particles), the pattern on the board behind will be similar to the slits, - two bright horizontal spots, but if the light passing through the slits is in the form of waves, the pattern on the board behind will show wave interference with unmistakably different patterns - (parallel strips) of light.

This experiment has been done may be millions of times, and the results are very consistent and unmistakable. The big shock from this experiment comes with the weird and absurd finding that the particles exist in two states –waves and particles. And the free particles can be in two or more places at the same time.

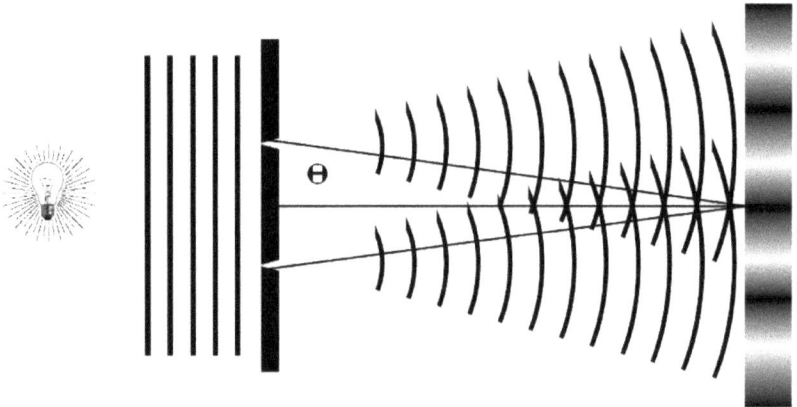

Double slit experiment

The two physical properties of particles – momentum and location – are in mysterious dependency. This is the next mystery - the more you learn of one of the properties of the particle - (position), the much more proportionately less you can know for the other property - (momentum). This dependency of knowledge appears to be the ultimate limit of our knowledge, and actually preventing any possibility ever (no matter how advanced we become, or how sophisticated computers we have) of predicting or arranging the future, because if you know the position and velocity of the particles or (the components) of some system, you can predict or arrange their future. And it is exactly this knowledge that is not allowed by the law of physics – to know and arrange the future!

We are given freedom of choice, but the outcome is uncertain and unpredictable!

The next surprise came when scientists learned that there is a mysterious informational link between pairs of particles – (entanglement). No matter how far they are from each other, the information between them travels in an instant. At present, the elite is trying to use this phenomenon to create a quantum computer, which actually is an attempt to use the intelligence embedded in the structure of matter and universe to create a super-computer.

Then in the experiments comes the strangest phenomenon of all time – consciousness! The scientists discover that consciousness affects the functions and properties of matter. No matter how cleverly they designed the experiments to avoid this phenomenon, it become absolutely clear that our consciousness knowledge affects the properties of particles and effectively collapses their wave functions, forcing them to crystallize like particles. Even a single fact of any kind of knowledge for the path (or position) of the particles always leads to the same result – a collapse of the wave functions. The puppet physicists call this the 'Observation Paradox.' For them, the fact that the universe is constructed with logic and is governed by intelligent order and consciousness is a paradox.

For me, it is very, very logical outcome.

From this point, we are bombarded with many divided opinions, strange theories, and explanations, but in general terms, the nonsense has prevailed. The puppets of the establishment avoid any studies and research on the fundamental properties of particles, forces, or laws of physical interactions. To cover up and give some sort of explanation for the relation of consciousness and matter, the sponsored teams come with clever solutions and state that 'the observer creates the reality', - a reality that does not exist! – A reality, which exists only in our mind! With aim of denouncing the meaning and significance of the new findings, the puppets with one statement just deny the existence of the entire universe! Can you believe such extreme arrogance?

They deny and don't want to tell you that the free-traveling particles possess all those strange and mystical properties described by quantum mechanics, but when these free particles bond with other particles and form solid matter, those properties are canceled, and the particles lose their individual properties and freedom. They obtain a strict location and become part of the collective order governed by the normal classical laws of physics!

The principles of quantum mechanics do not dominate the matter on a grand scale, because on this level the classical laws of physics taking over. - This is the fundamental element that the sponsored team absolutely unreasonably avoids, and are trying to pass the physical properties of free particles into solid matter. They claim that reality does not exist, or exists only in our minds. – Well done!

I will expose immediately the nonsense of this claim with this question: if the observer creates the reality, then who creates the observer and where does the observer come from and where is situated to create the reality? To have an observer, he must exist – and this is a reality! Einstein has told them, - 'I don't have to look up to know that the moon is there.' Even such a simple, obvious, and undeniable example has been ignored.

Later, to expose the nonsense of rejecting reality, Schrödinger created a clever hypothetical experiment. - A cat is locked in a box with a bottle of poison with a 50% chance of the poison being released. It is more than obvious that, even though we don't know what will happen in the box, the cat can be only in one of two states – dead or alive. This is an obvious and unquestionable example proving the nonsense of the Copenhagen Interpretation. The puppets have turned it into a circus and announce that this experiment presents a 'big mystery' and call it the 'Schrödinger Cat Paradox.' They claim that the cat is in a quantum state, and simultaneously is dead and alive. Are they mentally ill? What sort of a paradox is it? The cat is not in a quantum state, but is a real cat and can only be dead or alive, but not both together!

Schrödinger's (quantum) cat

From one side they carefully hid and covered up the newly discovered properties and order of the universe, and from another, they suppressed all attempts to study or find answers to simple fundamental questions, such as: what the spin is? - (the angular momentum of particles) and who determines and provides it? What actually determines the particular quanta (amount) energy of the photon? What determines when electrons to jump orbit? Why are the electrons in atoms not interfering with the external magnetic fields? Why are the atoms stable? Why is the proton weight such a weird number? What is the electromagnetic force? What are nuclear forces? What is gravity? What is time? What is space and what are the physical properties of space that enable it to act as a medium, making possible light waves to travel through space? (Because, if space is nothing, - this 'nothing' cannot propagate any waves.) And the biggest question: what is the information? Is the information being a manifestation of the hidden laws of physics or it is the actual laws of physics? What is the relation between information and consciousness? Why is the universe so precise and logically constructed? Quantum mechanics states that information cannot be created, cannot be lost, cannot be duplicated and cannot be destroyed. They hide the facts that information travels instantly to any distance and travels back in time.
I would like to clarify my information about this because this is very important for understanding the hidden functions and structure of the world.
The double-slit experiment has been done on a universal scale with light beams coming from a star 12 billion years away. The two beams of light passed on two sides of a massive object ('gravitational Lansing') and traveled separately for about 12 billion years. When scientists started measuring one of the beams, the other beam disappears. When they stop measuring, the other beam appears again. No matter how many times they have repeated this, the outcome always remains the same! That means that the present experiment forces the light waves (only the measured portion) from 12 billion years ago to pass only on one side of the massive object in the form of

photons. The two light beams have been separated for 12 billion years and the only possible scenario for this phenomenon is the information to go back in time to the point of separation of the beams and arrange exactly this portion of the light, which will be measured after 12 billion years to pass only on one side in the form of photons. - This is a real scientific experiment, and the findings and conclusions are beyond doubt! The other fact, - that the information is not affected by time is when the entangled particles are further apart, and one of them has significant speed, then those two particles are effectively at different times (according to the Theory of Relativity) but the information between those two particles is not affected by their time difference. This is solid confirmation for the phenomenal properties of information to travel instantly anywhere and back and forward in time. I just would like to explain the puzzle: is not the consciousness the one, who collapsing the wave function of particles, but the information does! - Our conscious knowledge is no more or less collecting information.

Now it is time for us to make a summary of the findings on quantum mechanics and to try to make sense of the available facts. Quantum mechanics has changed classical determinism and added the principles of uncertainty, free will and a limit of knowledge, where the individuals have consciousness, which gives them the ability to have free choice and to determine their own actions and possible future. The findings in the early 20th century in quantum mechanics have revealed a new understanding of the structure of the world, where the world is constructed mostly from non-physical elements. The physical element – 'matter' is our visible world described by classical physics. The non-physical, non-material elements of the world are time, space, information, consciousness, and the laws governing the universe. There are strict relations between all those elements in which are embedded unimaginably ingenious intelligence, precision, and logic, formed from one physical substance only: energy, - and from this energy comes the endless variety and beauty of the universe!

It easy to realize that consciousness and information are the actual laws of physics and that consciousness is the product of an intelligent mind. The major shocking realization was that the universe had existed long before us without the necessity of our consciousness, and the laws of quantum mechanics require a conscious observer on a universal scale for the universe to be formed and to exist and function in the way which it does. I don't have to explain to you that the only possible conscious observer on a universal scale is the intelligence, which we are usually called 'God.'

The scientists also found that our consciousness is related and is part of the universal consciousness but is much weaker. To a very limited extent, it can affect the properties of elementary particles. The founders of quantum mechanics found that deeper in the fabric of matter is embedded the principles for freedom of choice, <u>and there is also embedded the limit of knowledge</u> of how to predict, alter, copy or destroy matter and know the future of the universe, called - the 'Uncertainty Principle.' There are embedded a superfast informational link between all particles of the universe, but this information is protected from us with an absolutely unbreakable code. All these measures are logical and are created as a 'stupid–proof' structure of the world. These measures are an undeniable sign of a super-intelligence behind the scenes that protect the universe, protect the other civilizations and even us from obtaining knowledge on how to alter or destroy the structure of matter and universe. The properties of the universe are fine-tuned with incredible precision; scientists calculate the chance of this to be accidental is 1 to 10^{123}. This number is much bigger than the number of all the atoms of the visible universe - (10^{80}). Is there any logic in believing that the chance of 1 is more credible than that chance of $10^{123,}$ which is equal to many, many trillions?

The uncertainty principle in the micro-world is condition inserted only for us – to prevent the unethical society from interfering with the order of the universe. The proof for this is that from this apparent chaos and uncertainty emerging vast universe precisely balanced, driven by certainty and unbreakable physical law!

All these new findings reveal a new understanding for the origin and structure of the world, where the mind of the creator of the world exists as a part of the physical world and his presence and influence is embedded in the fundamental structures of matter, in living nature and even is part of us. Admiration is the term to use when describing the fundamental structure of matter and the laws of physics, where the principles for freedom of choice, ethical standards and limits for destructive knowledge on the grand and universal scale are presented.

The sophistication of the universe and its logical and unchangeable principles is in huge contrast with the row and chaotic behavior of the natural world. The natural world is not logical, is not economical, not ethical or perfect; nature is spontaneous, chaotic, cruel, and often destructive. It is sad to tell you such things that humanity has been robbed of the most valuable knowledge and ethical principles.

This knowledge was hidden and suppressed by the establishment. - The knowledge that will give people an understanding of where they come from, what values there are in life, and how to form an ethical, social order. The knowledge that has the power to unite the people under common principles of morals, ethics, and tolerance. - Principles are rejecting cruelty, corruption, lies, and tyranny. The knowledge that could guarantee the survival and prosperity of the human race and its future! The facts that our consciousness is part of the universal consciousness explain the sources for common human knowledge and beliefs in God and the origin of common principles of all religions. There are very fine hidden feelings for the mysteries of nature, which we normally suppress and ignore. This also explains the spiritual and emotional bonds between people who love each other. There are many conflicting statements about Einstein's beliefs. He never associated himself with any particular religion, but his logic, knowledge, and intelligence are undeniable, and I will insert here two of his statements. You can draw your own conclusions in what Einstein actually believed:

'Everyone who is seriously engaged in the pursuit of science becomes convinced that a spirit is manifest in the laws of the universe – a spirit vastly superior to that of man, and one in the face of which we with our modest powers must feel humble.'(A. Einstein)

'I want to know how God created this world. I am not interested in this or that phenomenon, in the spectrum of this or that element. I want to know His thoughts; the rest are details.'(A. Einstein)

We have to be careful. I would like to draw a clear line between scientific evidence, the laws of physics, and mysticism and deception. I strongly reject all those activities that are not based on scientific evidence and are designed to trick, confuse and manipulate people, such as mysticism, astrology, religious sects, witchcraft, ghost spirituality, alien abductions and any sort of miracles.
All that nonsense and activity is designed to confuse, manipulate, and use people for the wrong reasons, and together with orchestrated public misinformation of the pseudo-science are tools for deception. We just have to leave them behind and let knowledge, real science, and ethics to be our priority.

EINSTEIN INCORRECT ASSUMPTIONS, CRACKS IN THE STANDARD MODEL OF PHYSICS AND HIGGS BOSON NONSENSE

The Theory of Relativity and quantum mechanics are the fundamental basis for our understanding of the world and are the fundament of the so-called 'standard model' of physics. Unfortunately, they are incompatible, producing conflicting results and one of them is wrong! It is a well-known fact that both of them cannot explain the basic fundamental properties of matter and the universe.
The statement that this is our "best theory" cannot justify its lack of basic understanding and lack of any attempt in the last 100 years to study those basic fundamental properties of the physical world.
Despite the flood of wire mathematical fantasies, quantum mechanics is based on good experimental data and probably can be fixed. This is not the case with the Theory of Relativity, which basic assumptions are incorrect, and this makes impossible the theory to be corrected or saved.
Before start my consideration, I would like to inform the reader of the basic

physical principles on which our world is constructed, and they are: The material structure of our Universe (Space, Time, Matter) is made of a common form of energy. - You cannot create or dissipated energy! Energy is equal mass. There are no such things as negative energy, negative space, negative mass! Assumptions that going against those Universal physical principles are pure fantasy! There is no such thing as mass-less particles! Anything possessing energy, having mass, and the consequence of this principle is that the hypothetical mass-less particles cannot have energy! Without mass and energy, nothing can exist!

I will start my explanation with the fundamental elements of the material world - space, time, gravity, electromagnetism, nuclear forces, quantum information, and consciousness. Surprisingly, our "best model" refuses even to comment on the basic fundamental question – where the laws of physics come from and what mechanism these laws are using to enforce its physical order?

We brand ourselves as intelligent creatures, but the way we create our knowledge proves exactly the opposite. The adapted shameless direction of the so-called 'Copenhagen Interpretation' has done the job it was created do: To stall our progress and knowledge and to push research toward a dead end. The beginning of the 20[th]century was a time when we really had no idea about the fundamentals of physics and how they work. - And exactly at this time, the Copenhagen Interpretation stated that: "quantum mechanics had a complete description of the world and further study of the fundamental properties is not necessary"! - And this is the road we have followed for the last 100 years! --Avoiding at any cost finding the fundamental property of the world! As a result, we still have no idea what the physical properties of space are. We don't know what the time is. We have no idea how physical forces work to create attraction. We have no idea of the nature and constituencies of electromagnetic forces, no idea of the real nature of nuclear forces and matter decay mechanisms. We even ignore recognizing the role of quantum information, which is the actual informational link between particles of matter, and which is the physical law's enforcement tool. We are also ignoring the element of consciousness embedded in the structure of the physical world! - In this obvious lack of basic knowledge to say: '**Quantum mechanics have a complete description of the world, and further study of the fundamental property is not necessary!**' – Is supreme arrogance!

I am sorry to tell you that this mess and lack of basic knowledge is the real state and foundation of our present scientific knowledge. - When we have no real understanding of the fundamental principles and functions of matter and the world, it is absolutely impossible to have any credible further scientific advances. The incorrect basic assumptions are leading only to absurd theories the creation of mystical elements in the properties of the world with no connection to reality. Clear examples of this are bizarre new theories and developments in modern physics. –String Theory, holographic universe, multi-verse, and universe where 94% of matter is invisible and undetectable! - Those weird inventions become some sort of celebrated public fantasies, but without any proof, any credibility and no any experimental evidence in support.

Let us start by considering some aspects of the 'standard model' and its core - the Theory of Relativity. I have great respect to the author of the Theory of Relativity, Albert Einstein. His contribution to our knowledge is undeniable. He gives us the idea, that space, time, and matter are physical property related to each other. He gives us the famous formula $E=mc^2$, which telling us that matter is condensed energy. He did as much as was possible in his time with the then-current level of knowledge. We also have to be realistic, and cannot expect him to have given us all the answers and solutions of physics. His greatest contribution told us that the fundamental properties of the world are related to each other and the energy is the fundamental substance and building block for all constituencies of matter. - So far, these assumptions still have not been proven wrong!

Despite of his achievement, Einstein knew that there was something incomplete or fundamentally wrong in his assumptions. Contrary to other scientists, he had a critical and honest judgment of his own theory. In his latest years, within his close circle, he expressed his doubt and stated:

'You can imagine that I look back on my life's work with calm satisfaction, but from nearby, it looks quite different. There is not a single concept of which I am convinced that will stand firm, and I feel uncertain whether I am in general on the right track.'

Despite this, Einstein had no desire to confront the establishment, how Edwin Hubble did because they had blocked and silenced him.

None of the three classical tests proving his theory. They are:

1. The perihelion precession of Mercury's orbit. - It is exactly opposite than the predicted time dilation near massive body and in reality, is disproving the Theory of Relativity despite "their" opposite claim!
2. The gravitational Lansing, - which is caused mostly by light refraction of solar atmosphere, and not disproving at all that photons having mass!
3. The gravitational redshift of light, actually proving that gravity is a force, and photons having mass and are affected by the gravity!

In a letter to London Times 28 November 1919, Einstein state:

'The chief attraction of the theory lies in its logical completeness. If a single one of the conclusions drawn from it proves wrong, it must be given up; to modify it without destroying the whole structure seems to be impossible.'

And he is absolutely right! The real basis of his theory is built on non-scientific and incorrect physical principles. Scientific theories have to be correct in their basic principle assumptions and to have real proven physical foundations, where at most only one component is unknown, and the theory uses the other known phenomena to find the way to prove the nature of the unknown phenomenon. Unfortunately, the Theory of Relativity is constructed purely on unknown physical phenomena, where something we don't understand (gravity) is explained with something we don't know (space)! Such assumptions we call – 'speculations.' Unfortunately, this is the best description of our current knowledge and description of the world.

I would like to give an example for the absurd Einstein's assumption of the "curvature of space", which even person with basic knowledge of physics will understand: -Van Flandern state "Logically, the small particle at rest on a curved manifold would have no reason to end its rest unless a force acting on it"

In addition to his statement, I would like to give another example -

The law of physics postulates that moving objects have tend to travel in a straight line in space. If there is a curvature of space, the moving object must follow the path of the curved space, because this will represent its strait line of travel. Contrary to Einstein's claim, that the space is curved and this providing the gravitational attraction, the moving objects proving him wrong, by having escape velocity, because how I stated above, no matter of the mass or velocity of the traveling objects, they must travel in straight line in space, and if space is curved, they must follow this curvature regardless of their mass or velocity because this will be their straight line! Contrary to the claim, we observed that fast-moving objects passing near the centre of gravity have a tendency to continue travel in a straight line, where as much faster they travel, much more they are unaffected of the hypothetical "space curvature." The example of the membrane, with a heavy object in the center used to explain the hypothetical curvature mechanism of attraction, is more appropriate to be circus trick, than scientific explanation, because there is used the earth gravity to provide the effect of attraction.

I will suggest this experiment to be demonstrated to the public from space – from the International Space Station.

The next fundamental mistake of Einstein is that he incorporates together Space and Time in so-called "space-time." I agree that the physical components of the universe are related to each other, but Space and Time are separated physical phenomena! This wrong assumption still is preventing mainstream science from finding how Time works and how time providing the irreversibility of physical processes. (The explanation of Time puzzle you will find in a following chapter)

The first doubtful assumption of the Theory of Relativity is the curvature of space as an explanation for gravity. In fact, we don't know the real physical nature of space, and this prohibits us from making final and fundamental assumptions about the behavior and properties of space in order to explain the other unknown elements like gravity. – **You cannot explain something you don't know with something you don't understand!** (Val)

There is no any proof that space can be curved. The observations that gravity exists as a real attractive force is much more credible because space curvature is not proven at all, and lately we have found that photons have mass and momentum and inserting measurable pressure on surfaces! Einstein assumes that the celestial bodies are emitting force in order to curve the space. I am sorry, but if the matter emitting force what this force will be if is not the Gravity? Why should we go in such useless mental exercise to bend Space, assume that particles are mass-less and ignore the obvious? Space curvature is hopeless at providing an answer to the other three attractive forces. - There must be a common universal mechanism for the attractive forces which we currently don't know and don't understand how the attraction works. (Explanation of gravity mechanism and attraction is in a following chapter).

The next doubtful and really wrong assumption is that photons are mass-less. The Theory of Relativity is based on this assumption in the aim to prove that space is curved. But mass-less photons are simply - fantasy! It is another mystical assumption. – **'You cannot create something from nothing'** – as (mass-less particles). The photons are inserting measurable pressure on the surface! They have momentum, and to assume that when you accelerate "nothing" this "nothing" will have momentum is ridicule and stupid! Such an assumption is not scientific and does not even deserve consideration! - You cannot have a mystical and baseless foundation of our most important fundamental scientific theory! - Full stop!

The next incorrect assumption is that photon is the fastest traveling particle in the universe, but the latest observations are proving this assumption wrong! In the observed explosion of supernova 1987A, the neutrinos come to earth two hours earlier than the light! To cover it up "they" come with the explanation that the light has been trap inside of the exploding matter. I am sorry, but we have thousands of similar experiments with detonating atomic bombs, and the neutrinos and the light of these explosions coming out simultaneously! The length of the actual supernova explosion is about 10 seconds! Is not for two hours! How stupid is the explanation that the supernova explosion is not emitting light in the first two hours?

The scientists state that photons are mass-less particles. At the same time, scientists say that photons are packages of energy. - This is a conflicting statement because any form of energy is equal to mass! (For example, "they" are regarding the gravity of hypothetical Dark Energy as mass!) "They" admit that photons have energy and momentum, (which means that it has mass), but they never say - when the photons are mass-less? Why? - Because if you take all the energy of the photon, the photon does not exist anymore, but when it exist, it has energy, which is actual mass! Even the famous formula, $E=mc^2$, shows that anything possessing energy must have mass! And this undeniable fact is ignored by the Theory of Relativity in an aim to use the gravitational Lansing as proof for the curvature of space! The real **fact is that the Gravitational Lansing is not disproving the existing of attractive gravitational force at all! - Gravitational Lansing beyond doubt proving that gravity and mass of the photons are real!**

You cannot prove anything with wrong assumptions and by ignoring the facts!

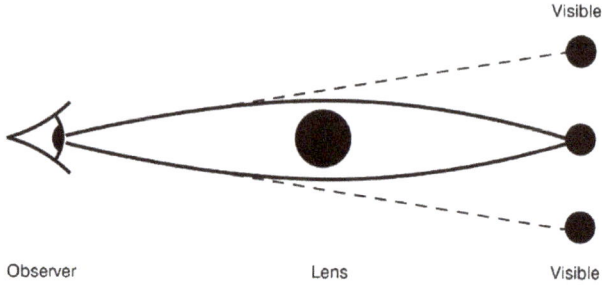

Visible

Observer   Lens   Visible

Gravitational Lansing

The fact that the photons possessing mass is enough to disprove the assumption for bending of space, but there is more:

Gravitational Lansing "proof" is also ignoring another three well-known facts, which produces the same effect of bending the light direction:

The first ignored effect is the light refraction - when light passes from the less dense medium into denser, it changing its direction (bending)! - The observed light from stars was passing through the solar atmosphere, which changes its direction! – It is simple! The sun's atmosphere is extending a few million kilometers above its surface and is causing the refraction of the passing light! (Same as the bending spoon in your cup of tea). Highly educated people ignoring such basics facts cannot be called scientists, but they are deserving more appropriate title as- "educated charlatans" – (see the image below)

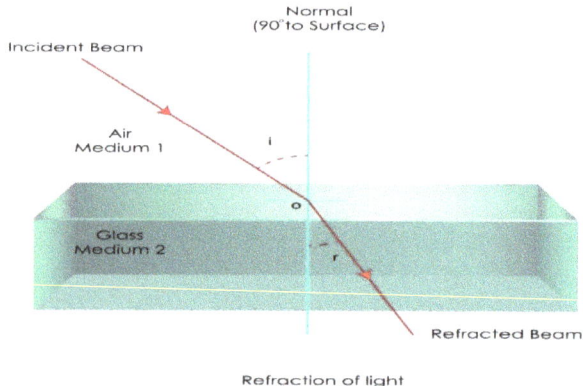

Normal
(90° to Surface)

Incident Beam

Air
Medium 1

i

o

Glass
Medium 2

r

Refracted Beam

Refraction of light

Science   **Light Refraction through Glass**

The next ignored fact is the relativistic time difference in the rotating sun atmosphere! – Everybody knows that GPS system has to compensate the GPS beam distortion and the error due to the difference in times pace between GPS satellites and earth surface. The observed light beam in gravitational Lansing passing true millions of kilometers solar atmosphere where each layer has different time speed, which will affect the speed and direction of the passing beam of light!

The next ignored effect is - the effect of the sun's magnetic field on the passing beam of light. Even high school kids know that the earth's magnetic field is deflecting the harmful sun radiation and protecting us! The earth ionosphere deflecting and bending radio waves! The solar corona is millions of kilometers tick ionized gases and is impossible not to have any effect on passing electromagnetic waves - (which the light is)

To ignore all those well-known facts and to accept the impossible and unproved scenario for mass-less photon and bent space is unwise, irresponsible, and arrogant! - In fact, the Theory of Relativity actually proves my statement that photons have mass with its prediction for gravitational red-shift–which is observed, proven, and well documented! It is obvious that the mass of photons interacts with the gravitational field of the massive object, or of universe gravitational field, loses energy, and appear red-shifted! - The best evidence is the effect on the light coming from a massive object like quasar incorporated with galaxy, where the quasar has significantly stronger light redshift than the attached to it galaxy - ('Seeing Red'- Halton Arp). The mathematicians do not understand physics, but I will try to explain in a very simple and clear way what the key is: - It is important to get to the bottom of this subject because the standard model is based on the incorrect assumption that space curvature provides gravity and the "best proof" for this is the gravitational Lansing of "mass-less photons". The fact that photons are possessing mass proves exactly the opposite -**that the gravity force, not the space changing the light direction together in combination with light refraction, relativistic time difference and magnetic field of the observed massive object!** - (Star, or galaxy)

I believe in the correctness of the assumption that all the properties of the universe are related and any form of energy is actual mass!

The explanation for the limit of light speed is simply that the physical property of the space is providing this limit! - Light behaves in a predictable way as a normal physical object with mass; and the speed of light is also affected by the property of the medium where it travels, - such as water, glass or air. Empty space cannot propagate electromagnetic waves – same as sound! To propagate waves, you need a medium, and space is acting as a physical medium. – (This is important)!

Let us consider the speed limit of something that we definitely know that is real, exists, and has no mass –quantum information. It is a real physical component of the matter! When information travels, its mass-less property prevents it from gaining kinetic energy. This means that the speed of information does not affect the energy balance of the system! And that's why it is not dependent even on time. - This is exactly what we observe. The information travels instantly from any distance, back and forward in time. The information is not time-related. - It is proven! You just have to know and understand the principle! And this is the correct explanation, based on the law of physics and our observations.

The other proof for the wrong assumption for mass-less photons is when they try to apply this assumption to calculate the well-known mass of the electron. What happens? - They got an infinite mass of the electron! If their assumption is correct, the results have to be accurate to +/−1%. The fact that the result is trillions of times incorrect can tell you the truth! The other proof that the photons have mass and momentum is the proposal for solar sails and photon propulsion engines to propel satellites faster than rocket engines.

The basic assumption of the Theory of Relativity is the formula $E=MC^2$, which is mass-energy equivalence. OK, based on this formula the mass of accelerating particle will gradually and proportionally increase with increasing the speed of the particle. And the particle to reach the speed of light must obtain infinite energy – or energy bigger than all energy of the universe! Let see if this is correct: Everybody knows that the particle accelerators like CERN accelerating charged particles to 99.999% of the speed of light. Even for non-specialist will be absolutely clear, that if the basic formula of the theory is correct, even a single particle accelerated to this speed must obtain the mass equal or greater the mass of our universe! And if this is happening the mass of the accelerated particles will rip off the foundation of the particle accelerator! Will destroy it in an instant! The fact that this is not happening is undeniable proof that the fundamental assumption of the theory of relativity, the basic formula and physical values of the standard model cannot be correct! According to their theory ($E = MC^2$) the CERN will need infinite energy to accelerate one particle to that speed, but this also is not happening! They are using just reasonable amount of energy to achieve this speed for millions of particles each second!- (How come?)

There are produced various calculations and "excuses" as "relativistic mass" and "rest mass" in order to deny the obvious facts, but those excuses are no more or less speculative adjustments of the property of the universe to fit their incorrect equation! To cover up this obvious crash of the Theory of Relativity, the pseudo-scientists have created also the 'renormalisation' – calculation, which strips the electron of its real properties, and uses the sign minus to insert into the poor particle an infinite amount of cancelling forces, (which are not more or less than "negative energy"). Nobody ever has heard of the existence of negative energy, but the crooks branded themselves with the status of "geniuses" for this! Awarded themselves with Nobel Prizes for this nonsense, and in this way, they have validated this great and arrogant deception.

As I stated, the people who understand physics adjusting their equation to the properties of the universe, but the mathematicians, who don't care for physics, adjust the properties of the universe to their equation. - This is exactly what they have done with the renormalization'. On the base of this mystical and wrong assumption, that "something" can be made from "nothing," the standard model has assumed that gravity force does not exist and all particles are mass-less! - This is the tragedy of the weird foundation of our current "best knowledge." The chain of fantasy and incorrect assumptions start piling up, and this creates the necessity to insert into those fantasies more and more nonsense like some sort of gravity mechanisms for the "mass-less" particles. (Because according to pseudo-science all particles are mass-less). According to the law of Physics, something which has no mass cannot have energy- (or cannot exist). The creation of hypothetical Graviton and Gluon was not enough to explain these incorrect assumptions and dictated the creation of new hypothetical Higgs Field and Higgs Boson, or the "God particle." The incorrect assumptions continue piling because such universal fields as Higgs Field have to be anchored somewhere to resist the movement of the massive universal structures, (because every action have the opposite reaction) The existence of such enormous fields needs enormous opposing energy, whose origin is not known, is not proven to exist and creates problems with the conservation of energy and the energy balance of the universe. To explain how particles, acquire mass, mathematicians start creating separate complicated mathematical theories for many different particles.

Anyway, they did not explain how the Higgs Field interacts with the wave state of particles because it is assumed that most of the time the particles are in the form of waves - It is a well-known fact that the interaction with waves is possible only with waves, and these waves must have the same origin and physical properties and must be in phase with each other. The particles have millions of different frequencies, and we know that when waves interact with each other, they gain or lose energy, nothing else! Do you realize the magnitude of the absurdity? - The Higgs Field must appear in a different form of waves for each particle and to be virtually the same as each of them! - This is a creation of another hypothetical parallel universe inside ours, but still, even this does not give mass to anything!

They don't even explain what the physical meaning of 'field' is! What are its constituencies, its state, origin, and intensity! **The only evidence provided for this bizarre theory is some insignificant bump in the graph of particle decay.** How possible can this bump explain the entire phenomena of gravity? - Such a bump could be an energy spike for any other reason! (The best details for this shameless scam you can find in the book, '*The Higgs Fake*' by Alexander Unzicker.)

The standard model is also assuming that the speed of gravity is equal to the speed of light. This assumption also could be incorrect, because the gravitational force can exceed greatly the speed of light. The fact that gravity is physical force dictate that the speed of gravity will be finite, but can be any speed!

The fact is that the standard model explains the universe based on gravity and ignores electromagnetism whose strength is many billion times stronger than gravity - (10^{38}), can tell you that something very, very wrong is embedded in this model. - **This model is based on gravity only when the normal visible matter is only 1%. The majority of matter in the universe is 99% plasma, and plasma is ruled by electromagnetism, which is many billion times stronger than gravity!** (I am not considering the "dark staff" at all)

The atoms and matter are held together by electromagnetic forces. All chemical reactions have electromagnetic mechanisms, all the life is supported by the photo-synthesis, our brain using electrical impulses, and we just ignore to consider this superior-in-strength and versatility universal force. Why? - To maintain the inserted dogma and nonsense? What is the reason to continue deceiving ourselves?

We haven't got basic knowledge even for the structure of atoms! – The assumption that the strong nuclear force is property and product of the nuclear particles and is situated evenly between them is not correct, because, in this scenario, the atomic nucleus can grow indefinitely and will be stable at any size! Fact is that the nucleus of the heavy element becoming unstable with an increased number of nucleus particles. **This fact proving that the strong nuclear force is situated in the center of the atomic nucleus and have a limited range from the center out.**

We don't understand even how our sun works! The claim of leading scientists that they know how the sun produces energy is absolutely false! The fact is that hundreds of billions of tax-payer money have been spent for research; the "required" conditions for nuclear fusion have been achieved, but nothing works! Everything has been tested for five decades, hundreds of billions wasted, and the chain of negative results is actual proof that those claims are wrong! - If the sun's heat is produced by the thermo-nuclear reaction, why doesn't the sun explode? There is nothing to stop a potential thermo-nuclear explosion. How the "mass-less" photons can provide "pressure" to counter-balance the enormous gravitational pressure of the sun mass not to collapse? Why the temperature of Sun is on the reverse sequence to the proposed model?- This is only a small example of the real pile of obvious miss-assumptions, and unexplainable phenomenon, which could be explained with electromagnetism, but not with gravity, bending space or the 'standard models.' It is obvious that the mathematicians don't understand that the solar corona is the place, where two electromagnetic fields meet and cancel each other, and the result is the heat radiation - similar to the arc welder or microwave oven, which we are using every day!

The universe is a dynamic system of matter and energy exchange, where the motion and rotation of the universal bodies creating interaction

between their magnetic fields and the magnetic field of the universe! This interaction creates currents, electrostatic charged regions, polarity, ionization, and in the conditions of superconductivity transferring and distributing mass and energy on a big scale and enormous distances. The scientists using magnetic fields to accelerate charged particles in the particle accelerators, but for the inserted mathematicians are "big puzzle"- what is "this," which accelerates continuously the charged particles emitted by the Sun! For them is a "big puzzle" also the basic understanding, that the interaction between the magnetic field of spinning star and the magnetic field of the galaxy producing electric current! For them is "puzzle" why the Sun is changing its Polarity every 11 years! There is also the interaction between the magnetic field of the rotating galaxies and the magnetic field of the universe — the aging stars are losing an enormous amount of mass in the form of ionized particles. The produced current of these interactions in the conditions is the driving force behind the plasma flow between stars and galaxies, and is a dynamic system of energy and matter re-distribution for the recycling system of the universal structures!

The recycling mechanism of the universe is the delicate dynamic balance between the saturation of the interstellar space with the released energy from the fusion of hydrogen and the rate of electromagnetic fission of the heavy elements back into hydrogen. - This is the balance between fusion and fission provided by the two long-range forces – gravity and electromagnetism.

Scientific research is stubbornly directed to study only nuclear fusion, and the electrical nature of these processes is ignored. They are claiming also, that electron is a fundamental particle, which means that we cannot create, or destroy electrons, but in simple Cathode-ray tube - (old TV screen) their "knowledge" is evaporating, because there the electrons are created by heating the cathode, and then the beam of electrons ended in the screen in order to be transformed into the light images, which we are watching. Same thing we observing with the light blobs - the heated tungsten coil emitting constantly photons! - Are we have to believe "Them", that we cannot produce elementary particles of matter without producing antimatter?

A well-known fact is that some scientists and electrical engineers, who announced that they have broken through to free energy from the earth's magnetic field, have mysteriously disappeared or have had accidents just before revealing their secrets. Ask yourself, why were all of Nicola Tesla's papers, when he died, been confiscated by US authority and was kept a secret from the public? - To promote science, or hide his achievements? The standard model is obviously based on the wrong fundamental assumptions and cannot answer even simple basic questions, but it is the most comfortable theory for the establishment and is promoted as a "triumph of our intelligence."

It is obvious that time is energy-related. The implication of the accepted understanding that the universe is a closed physical system, where space, time, and matter are related, leading to a simple and important conclusion! – Space is not expanding! Because if it is expanding, this will change the value of the other two components of the system – matter and time.

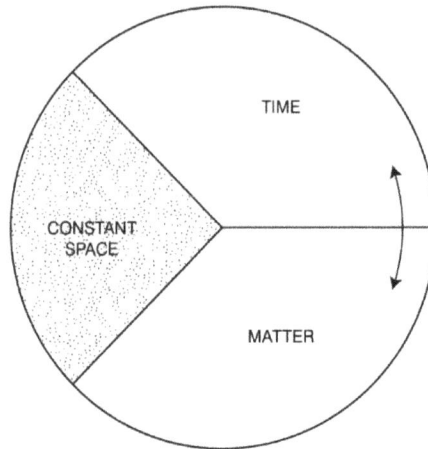

We have very good observational data for the past 13 billion years, and our observation shows that the amount of matter in the universe and the pace of time have been constant in this period. And when these two components of the closed physical system – (the universe) are constant, the law of mathematic telling us that the third component cannot vary! – It is constant too! This is simple and obvious facts, which can be understood even form a person without any training and qualification in astronomy and mathematics! If the scenario of expanding space is correct, we will observe that the time gradually will slow down, and the matter also gradually will diminish and disappear! (Space + Matter + Time = Constant). Fortunately, we are not observing such a catastrophic scenario. We know that Time and Matter is not varying!

This undeniable fact is ruling out any speculation for the space expansion! The tragedy is that the peoples who understand physics have been muted. The mathematicians who don't understand physics and adjust the properties of the universe to fit their equation have prevailed over logic and reason, and the result is obvious – a total scientific mess, where a leading scientist - (Richard Feynman) states: '**If somebody claims that he understands quantum mechanics, I will call him a liar**'. - And this is the best description of our current knowledge which is based on non-existing phenomena, mathematical fantasies, and weird assumptions that make no sense and have nothing to do with the reality of the world! The greatest tragedy of humanity is that the leading intellectuals have betrayed their own people and have joined the financial elite and their orchestrated campaign to deceive the World just for comfort and money.

The light refraction is not bent space, and Einstein knew it! The biggest problem for humanity is the fact, that the leading intellectuals have betrayed us and have joined the orchestrated campaign of public disinformation

Albert Einstein

'You can imagine that I look back on my life's work with calm satisfaction, but from nearby, it looks quite different. There is not a single concept of which I am convinced that will stand firm, and I feel uncertain whether I am in general on the right track.' (Albert Einstein).

The vital question – where the universe comes from – is one of the most important questions in our lifetime. The major problem is that the answer to this question has been carefully guarded by the elite.

The scientific standard is based on observations, experiments, and laws of physics, logic, calculations, statistical data and analysis of the facts - (All the facts). To proclaim that something is correct, the scientific result must be accurate, logical, and able to be repeated; To be accepted, the scientific results and conclusions must obey the laws of physics and to have accuracy of the experimental data at least within five decimal points. In science and in scientific theories are not allowed to be included any supernatural and mystical elements, or assumptions that violating the law of physics.

Let us see what kind of explanation is the present official answer to our vital question for the origin of the universe - The Big Bang.

The Big Bang theory is the dominant theory widely accepted by most people and mainstream scientists. It is accepted as a standard scientific model of our understanding for the origin of the Universe, but the Big Bang theory is just Theory, is not a fact! This theory is not the product of scientists, but of Belgium priest, and priests' usual practice is to use mystical and supernatural elements to convince the public to believe in his statements. - That's why the fundamental concept of Big Bang theory is mostly mystical and supernatural and cannot be qualified to be scientific theory!

1. The theory starts with the unscientific mystical assumption that the universe appears spontaneously from nothing for no apparent physical reason! - This statement is violating the fundamental law of physics for the conservation of energy – (you cannot produce energy from nothing!)

2. The assumption that the Universe starts from a singularity is mystical because the law of physics is not allow compressing the entire universe in space smaller than an atom! - (Plank Constant)

3. The assumption that Big Bang explosion can create the Law of Physics and the incredibly sophisticated plan of the Universe in a time shorter than 10^{-38} of a second is not even mystical, but is absurd and stupid assumption! (The explosions are not creating order, but disorder!)

4. The assumption that the Universe has expanded faster than light is also a supernatural element and violation of the law of Physics!

5. The introduced invisible, undetectable Dark Matter in the model is another mystical element involved to patch up the model's conflicting claims of the expansion ratio, size, and age of celestial structures.

6. To explain the miss-match of data and observations, 'they' invented the next mystical element! – The undetectable Dark Force.

7. The appearance of the visible Cosmic Microwave Background

Radiation from the opposite direction of Big Bang and a huge unexplained distance of 13.7 billion light years is absolutely miraculous! - (See the graph pg. 54)

8. The claim for the continuous glow of cosmic vacuum for 13.7by is miraculous and conflicting statements also!

9. The "formation" of a mature universe and second-generation stars in only 400m. Years is an absolutely miraculous statement!

10. The time of expansion and the size of the visible universe is the same! – 13.7 billion years and 13.7 billion light-years distance, which means that the expansion rate is equal to the speed of light. - And to see something retreating from us with the speed of light is absolutely miraculous claim!

11. The disappearance of antimatter is another mystical assumption!

Mystical and supernatural assumptions are religious tools, not scientific! Any self-respected scientists should not even mention them!
The mystical-religious explanation for the origin of the Universe is more credible than the Big Bang theory because at least it is providing a creator!
The Big Bang theory has no bases even to be qualified as theory! Because is ignoring facts, observational data, violate the law of physics and all its statements are a bunch of conflicting claims, which makes it impossible to be created a normal working model of gradually expanding and developing Universe and not to be in conflict of the observational data! That's why all models and graphs are presented in reverse sequence of the proposed Big Bang, in order the public not to figure out the impossible scenario and the miss-match of data, observations and conflicting claims of this mystical theory!
Is time to put an end to this orchestrated deception, because this deception is acting like a brick wall for the advance of science and our understanding of the world.
It is forcing scientists to study none existing phenomenon and to bombard us with bizarre and absurd findings, theories, and discoveries which have nothing to do with common sense, the law of physics and reality!
We will consider carefully every detail of this 'Story' to remove any doubt about the validity of this great deception!

<u>Here is a brief history of the Big Bang Theory:</u>

The first people to notice the receding of the nebulae were Vesto Slipher and Carl Wirtz in 1910. Albert Einstein came to the conclusion that the static model of the universe is not a stable model and will collapse. He introduced the cosmological constant to restore the gravitational balance of the universe. The Big Bang Theory was proposed by Belgian priest Georges Lemaître in 1927 on the basis of the observable red-shift (receding speed) of the nebulae.

In 1927 Edwin Hubble confirmed the red-shift and, together with Milton Humason, formulated 'Hubble's Law,' mathematical formula for the receding speed of the galaxies in relation to the red-shift. The conclusion of their observations was that all galaxies are receding at speed proportional to their distance from us.

(Later, when Hubble became familiar with the Theory of Relativity and the gravitation red-shift of the light, he denounced his assumption for the expansion of the universe, but this time he was just ignored!)

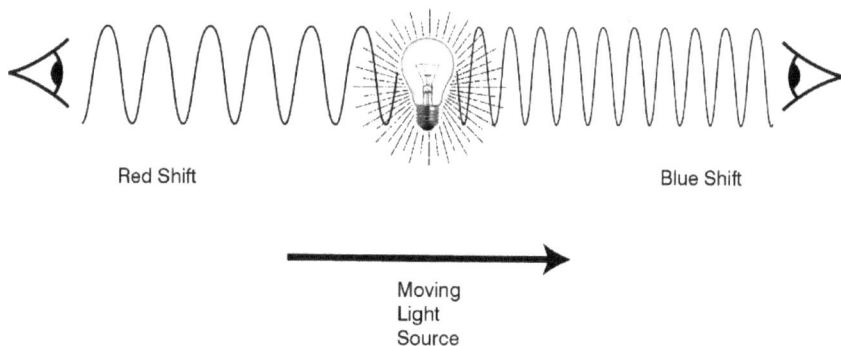

Red Shift   Blue Shift

Moving
Light
Source

Light redshift effect of moving source

In 1933 in his tour to the US, Einstein was accompanied by Georges Lemaître. Einstein was desperate to emigrate because of Jewish persecution in Europe. Under pressure, in order to emigrate, Einstein found support in the Belgian Catholic priest Georges Lemaître. What the deal was made we will never learn, but the fact is that Albert Einstein discontinued his support for the 'constant universe model' and withdrew his cosmological constant. Einstein officially recognized the 'correctness' of the proposal of Lemaître, and the Big Bang Theory was established. On the other hand, Einstein never denounced his findings for the gravitational red-shift of light! - And this is a reviling fact for his real position!

Gravitation Lansing and the claim that the gravity of black holes attracts and traps light is actual proof for the correctness of the fact that the gravity of the universe affects and depletes the energy of light and producing the light red-shift. The light red-shift cannot be used to prove that all galaxies are retreating from the earth. We know that the earth is insignificant and cannot be the center of the universe.

To solve this problem, in about 1980 the American physicist Alan Guth proposed that the explosive nature of the Big Bang be replaced with a uniform expansion. In a stroke of mental degradation, this guy went from the land of mysticism to the land of fantasy. To justify their scientific impotence, 'they' invented phenomena's as 'false vacuum,' 'magnet monopoles,' 'supercooled phase of inflation,' and the most important invention, - 'spontaneous inflation faster than light.' We have started our consideration with the laws of physics, that nothing can exceed the speed of light, and even if a single atom to do this, it will need more energy than the energy of the entire universe! How can Guth explain his assumption? Where is that unlimited energy to propel the entire universe faster than light to come from, and after that, where this energy has gone? A scientific theory cannot make such baseless claims. This is a farce, not science!

The point of validation for this new proposed super-fast universal expansion is the assumed "undeniable correctness" of Big Bang theory, which not living us any other option, then to accept this great violation of common sense and law of physics!

The model presented to the public looks nice, easy to understand, and looks elegant and logical. The theory explains that, when we look at the sky, we can see it as a timeframe of past events. The further we look, the further back in time we look, and the last thing we can see is the Cosmic Microwave Background (CMB), which is the remaining glow of the Big Bang cooled down to 2.725 Kelvin. A fascinating and nice story, isn't it? But it is real and correct? Before we start analyzing the model, I would like to draw your attention to the first two absurd claims, which are fundamental to the model. The claim, that when we are looking further away, we are looking back in time, is correct because the light needs 13.7 billion years of traveling time to come to us from the furthest currently observed object, - the CMB. But for some mysterious reason, the creators of the model forget that this object needs a minimum 13.7 billion years (if traveling at the speed of light) to distances from us to the present position, before sending the light toward us from a distance of 13.7 billion years. The sum of those two figures is 27.4 billion years! But the model claims that the universe is only 13.7-billion-years-old. It ignores (or just forgets) that the objects need many billion years to recede to such distance from us before sending us the light! (Or in short) - the light of CMB starts traveling toward us from 13.7 billion light years distance in the time when the universe supposed to be smaller than an atom and be situated in our point of observation! - This is undeniable observational proof that the size of the universe 13.7by ago has been the same as now! Are they really that naïve, or do they want to treat us like stupid?

The next obviously absurd (or deceptive) claim is that we actually see the present temperature of the CMB. We cannot see the present condition of CMB, because the light of the proposed CMB has to travel away from us and should be gone in the empty space! The visible CMB is coming from the opposite direction of the proposed glow of Big Bang! It is an obvious and simple fact!

In our analysis, such obvious misconceptions will be avoided, and we will build the model in normal sequences and consider everything in the correct order. Let us continue with the events surrounding the development of the Big Bang

Theory. Some mainstream scientists were 'concerned' that the temperature of the CMB is very uniform. This is not natural, because the heat information or (spreading of temperature) has not enough time to reach such uniformity over big distances. This is a very small and insignificant problem called the 'horizon problem.' This **'small anomaly'** is accepted as a fact and is left there unresolved. Actually, to solve this problem, an unnatural and mystical solution was proposed, - the initial universe expansion (faster than light) mentioned above. The problem in science is that there are no such things as small or insignificant anomalies because even a small anomaly tells us that there is something wrong with our assumptions or our model, but in this model, we are facing numerous huge mystical anomalies and hiding of 13.7 billion years expansion time. From recent history, we have many lessons where even very insignificant anomalies lead to the abandonment of a concept or hull theory. This is one example:

- The Geo-Centric Theory has been well established as very logical and as an official theory of the scientific world. There was only one very **'insignificant'** anomaly, where five stars were changing their positions– (the solar planets) and the theory couldn't explain this phenomenon. They just accepted it as a fact without explanation. It turned out later that this 'insignificant' anomaly was telling us that the whole model was wrong! Time has proved that this small anomaly was the revealing clue for the incorrectness of the Geo-Centric Model.

There is no 'small anomaly' in science! - Every detail of a new theory must be absolutely correct, proven, testable, and logical and must not violate the laws of physics. - This is the scientific standard for acceptance of a theory. Unfortunately, this is not the case with the Big Bang Theory!

If we would like to insert unproven, mystical or supernatural properties and functions, and start to explain scientific models in a mystical way and to ignore the law of physics, it will be better for us not to deal with science, but simply go to the local church or monastery, where supernatural events are well accepted.

I would like to help you become more familiar with some physical laws and the fundamental properties of the universe to be able to follow the argument with confidence and understanding.

From the Theory of Relativity, we know that the fundamental properties of the universe (space, time and matter) are related; altogether, they form one closed physical system where the laws of physics are not allowed to add or take away even one particle or atom. That means their amounts are in strict relationship with each other, and the total amount of energy (or mass) is constant and unchangeable! In short, you cannot change the amount of energy of the universe by playing smart games and disobeying the laws of physics in order to prove your theory!– (The claim for faster than initial light expansion)

The speed of light has an accepted limit, and no material object can exceed the speed of light, because the speed is kinetic energy, and even a single atom, to reach the speed of light, must obtain unlimited energy, that exceed the energy of the entire universe! And if this happens, it will cannibalize and destroy the universe because, as we stated above, the universe is a closed physical system and you cannot add or take away any energy. –Full stop!

The next thing, which is good to know is that the speed of light is constant in space and is not affected by the speed of its source. The consequence of this property of light is the red or blue shift. The light on the front of the moving source is blue-shifted (has more energy). The light behind the moving source is red-shifted (depleted of some energy).

When the hypothetical light source reaches the speed of light, it becomes invisible for the observer behind the fast-moving source, because the speed of the source extracts all the energy of the light behind, and this light effectively does not exist anymore. In short, you cannot see an object retreating from you at the speed of light! Remember this! - It is important for our further consideration of the claims of the Theory!

The elite succeeded in instituting the Big Bang Theory as the official scientific explanation for the origin of our universe. Everything had been quiet for a while, but advances in science and astronomy started to see some cracks in the validity of this theory. The first one came when the existence of antimatter was discovered. All the experiments consistently, without any exception, proved that when energy is condensing and transforming into the matter, it always produces the same amount of matter and antimatter. This discovery immediately raises a big question in the validity of the Big Bang Theory, because the produced matter and antimatter annihilate each other and return back to energy and light! So... how the light stage of Big Bang turns into the ordinary matter?

This problem was eliminated by the pseudo-scientists when, in the statistical data of an experiment, an error appeared, and one particle from billions wasn't recorded. This statistical error was proclaimed as a triumph of science, and this error "saved" the validity of the model again. But the problem for the model didn't stop there. The problems were continuing to come with the newly available data from astronomy. Hubble deep field photos show that, where there should be only developing primordial universe in the form of gas and dust (according to the model), there is actually a fully developed and mature universe! What to do now? The establishment starts to worry, but why should they? Our guys are very clever. They immediately invented some sort of mystical superglue or 'steroids' - (dark matter) to force this primordial gas and dusts to form the universe in a very short unreasonable time. Brilliant! - We cannot see or detect this matter, but we have to believe the smart guys that this illusion is there because otherwise, their model will crumble! But the cracks didn't stop coming! Ordinary scientists are not allowed to use the correct formula for the red-shift of light, which including the gravitational red-shift, the astronomers are using only what they are allowed to use – just

Hubble constant, and as result of this they "calculate" some fake receding speed of the galaxies. The new shock of this incorrect calculation came when we were able to see much further. The wrong calculations suggest that the universe is expanding with ever-increasing speed!- (Bravo, well done!)

What to do now? No problem! The guys are smart! They have invented the next mysterious substance – 'Dark Force' - (the next invisible, undetectable substance of the Universe), which pushes all the galaxies away with "negative' gravity and at ever-increasing speed! We cannot see or detect this new mysterious substance, but we have to believe the smart guys that it is there! Other ways their theory will be invalid, - which is impossible, because "They" are "etalon of correctness"!

<u>This is the main claims of the official Big Bang Theory:</u>

- The Universe comes out from nothing and for no apparent physical reason.
- When we look at space, we see past events as a timeframe of the development of the universe, and the last thing we see is the glow of the Cosmic Microwave Background (CMB) of the Big Bang.
- The universe starts with the Big Bang and is expanding.
- In the first 7 billion years, the expansion was decelerating; after 7 billion years, the expansion started accelerating at an ever-increasing rate.
- The age of the universe is 13.7 billion years old.
- The size of the observable universe is 13.7 billion light years radius.
- The universe is homogeneous and uniform, isotropic in all directions and has been homogeneous in all stages of its development from the Big Bang until now.
- In the time of recombination, the temperature of the hot gases has been about 3,800K.
- The CMB is the picture of the universe in the exact time of recombination – 380,000 years after the Big Bang – because it is very homogeneous. (In later stages this homogeneous gas become lumpier and starts condensing into stars and galaxies, and the space between them become empty, transparent and cold).
- The observable CMB is 13.7billion light years away.
- The glowing shell of the Big Bang has continuously cooled down and retreated to the present position of 13.7billion light years distance.
- The present temperature of CMB is 2.725K. (They do not specify the subject – it is the emission or the temperature of the "glowing shell")
- The present location of the CMB is 46 billion light-years away and is beyond the observable universe.

Let us see what kind of explanation providing the Big Bang Theory, which is the dominant theory widely accepted by most people. It is accepted as a standard scientific model of our understanding for the World.

<u>Here are ten points for consideration:</u>

1. What are we actually seeing? When was the proposed CMB has emitt such a uniform pattern and how hot supposed to be the CMB in this time? How quickly did the CMB become cool and transform itself into stars and galaxies, and are there any remains of its glow in stages as (1 billion, 2 billion, 3 billion and on…). And why are we not seeing this "retreating" shell of CMB in a visible light spectrum?
2. The current explanation of the CMB contradicts the basic principle of the model that the universe is homogeneous and uniform because it states that the CMB is a retreating glowing shell away from us. But does this glowing shell don't suppose to turn into stars and galaxies? Have any glowing shells been observed in any other stage of the Universe "development"?
3. At the time of emission of the proposed CMB, our point of observation has been inside in the CMB. This fact makes it impossible for us to have any chance later of seeing this emission because our point of space has been inside in the volume of CMB. The light of CMB must go away from us! And also - why are the galaxies between us and the visible CMB? - The galaxies supposed to be formed after the emission of CMB and they should be behind the proposed CMB! But the reality is exactly the opposite! - The galaxies are in front of the visible CMB! In NASA Big Bang model, our observation point is on the opposite side! - And why are no stars behind us in this model? How we get there before the light of Big Bang get there?
4. What can glow and what cannot glow? (validity of the glowing shells)
5. The distance between us and the visible CMB is 13.7 billion light years. To be on this distance in 13.7 by time, it must retreat with the speed of light which make it practically invisible! Why can we see it clearly?
6. According to the official theory, the CMB is now on 46 billion light years, what actually is its retreating speed?
7. To be on 46 distance, CMB has traveled only 13.7 billion years to get there! No physical object can travel three times faster than light! No physical object can be visible if it retreats at speed three times faster than light. How can the CMB be in same times at 13.7 and 46 billion years away? - To accept this, we have to violate the laws of physics, start believes in miracles or become incredible stupid!
8. The glowing shell of Big Bang has gradually cooled to 2.725 Kelvin, and that is what we are observing now. - This is an absurd claim because we can observe only the conditions of 13.7 billion years ago, not present-day conditions! And the gas this time supposes to be 5000

degrees, not 2.725K. - (The original calculation of microwave emission from Big Bang has been 50K) - The temperature of the observed emission is out with 18 orders of magnitude!   So...only the inaccuracy of 1,800% is ruling out the validity of this claim.

9. Why is there such a huge difference in the observed temperatures between the proposed hot gas of the 'Big Bang' – (the visible CMB) and the first stars, which supposed to be formed just a few hundred thousand years later? They must have had a similar temperature! We can see those stars in the normal light spectrum, but the visible CMB is 5,000 degrees cooler, and the model claims that this is the present temperature, despite that we cannot see the present temperature of object 13.7bly away!

10. The puzzle of dark flow – what causes this movement of the galaxies to move beyond the horizon towards the CMB? And what actually is the real nature of the observable CMB?

We have to be familiar and consider some aspects and physical laws which will apply to our consideration of the Big Bang Theory:

* Our universe is flat (the space), and space is uniform in all directions. Light travels in a straight line with a constant speed in all directions. (That means that we do not observe any expansion of space).
* The distance between us (now in present time) and the CMB is 13.7 billion light years. That means that the Big Bang is '0' time and the present time - 13.7billion years after this event and is in our location!
* The time for light to travel from the CMB to us in present time is 13.7 billion years.
* There are three separate factors for the red-shift appearance of the traveling light:
  a) A retreating source of light from the observer, stretching the light and cause red-shift. (Doppler Effect)
  b) The gravitation red-shift of the light. According to the Theory of Relativity, gravity depletes the energy of passing light, and this causes the red-shift, - (which was mistaken by Hubble as retreating of the galaxies).
  c) The electromagnetic field of the universe also depleting the passing light energy, because the light is electromagnetic waves!
* No physical object can travel at the speed of light or faster than light.
* Physical objects retreating from the observer at the speed of light or greater are invisible to the observer.
* Light cannot exceed the speed of light.
* Light cannot be stopped still or preserved in space to wait there for us to see it later – (after 13.7by – CMB)
* Experimental data shows that when energy condenses into matter

(E=mc2), equal numbers of ordinary matter and antimatter are always produced, with no exceptions.

- The claim that universe (or space) has expanded faster than light was added around 1980 without any proof and is done only to validate some unexplainable anomaly of the newly proposed Expansion Theory. It is a baseless mystical, supernatural assumption.
- The claim for the existence of dark matter, which is invisible and undetectable, is invented to prove the impossibility of a fast-maturing early universe. This is an unscientific, supernatural, and mystical claim, which does not deserve even to be considered.
- The next such supernatural (mystical) claim is the invention of the invisible and undetectable dark force in aim to prove the false assumption that the universe is expanding. This is another baseless supernatural claim, not deserving attention.

Let see what is the official explanation and the most famous picture of this event – the NASA diagram, or so-called NASA 'Big Bang Cup' It is easy to follow the proposed events from left to right, but if you start analyzing the claims and what is there, you will discover shocking discrepancies!
(See the diagram below)

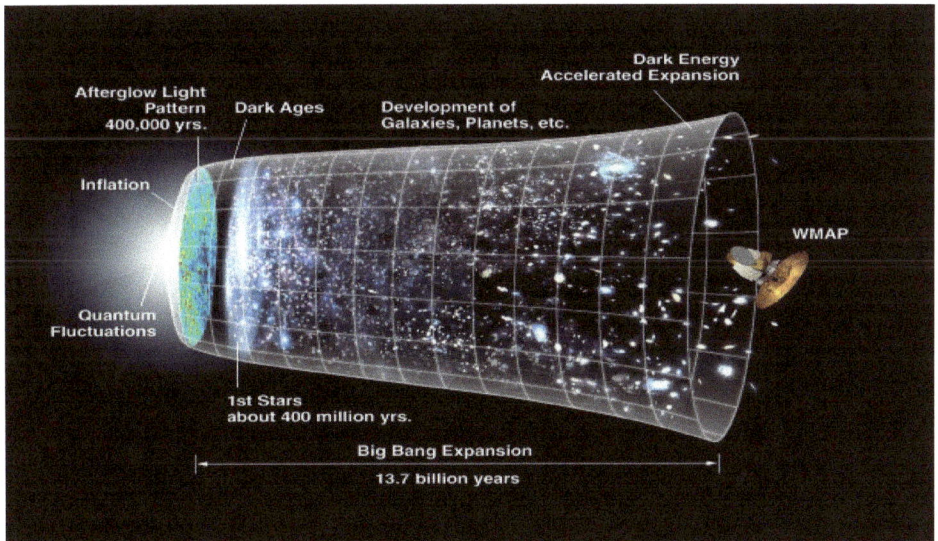

This is the famous NASA Big Bang model, but what really is representing, and why our observation point is on the right side, where is no stars behind?

How we arrive there before the light of Big Bang reach this point? They claim, that in present CMB is 43 billion years away, but this is nonsense, because in this case CMB will be at 32 billion years in <u>negative time</u>, - before BB and the expansion suppose to be three times faster than speed of light!

48

This is how NASA has constructed their famous (Cup) – the model of Big Bang: Instead of drawing realistic circular (cut section) of the proposed spherical universe, they create a mathematical graph. They entered the expansion rate in the graph (as the example below), where the top (black) side representing the expansion in time. In order to create confusion, they created the symmetric bottom (red) part, which is the nonsense of negative inflation. Then they have removed the graph arrows, populated the mathematical graph with stars and galaxies, then they switch the position of our observation point from the left to the opposite side on the right side and 'Voila!' we have the NASA famous cup of Big Bang!

As a result, NASA model becomes something which the people are pretending to understand, but they cannot! And where behind 'our' right-sided observation point (by NASA) is the absurd scenario of dark and empty space, with no stars! - This is absurd, which just confusing, but not explaining anything! – See the model of NASA (above) and notice in the lower graph (below), where is the correct position of our observation point and that the real visible CMB coming from the right side, (toward, not from Big Bang out) which is ruling out any possibility Big Bang theory to be correct!

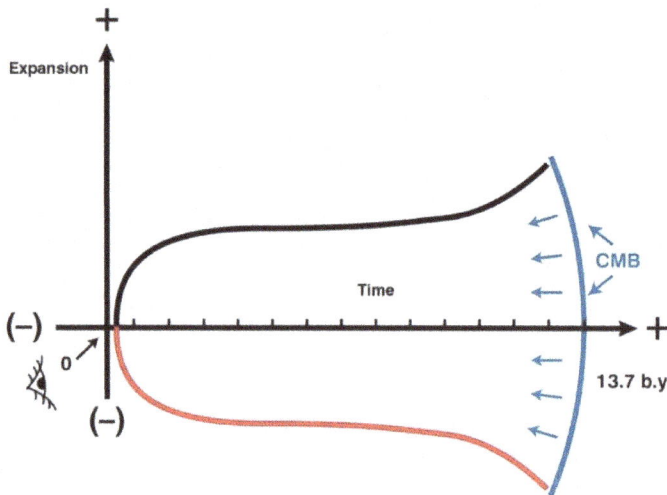

Here are the correct positions of our observation point and CMB correct position, and its correct direction of traveling.

We have to build the correct picture of the model from the beginning to the present time in normal sequences, not in the opposite direction. The only correct statement in the theory is the introductory statement, that when we look further in the universe, we see the timeframe of past events, because of the further we look, the further back in time we are looking. - This is because the light from the distant object needs millions or billions of years of traveling time to reach us. This is just an ordinary calculation of how much time light needs to travel to reach us. The problem is there that this statement totally disproves the theory of BB! We are observing a steady state of the universe; where behind the nearest galaxy is different galactic structure and totally different universe!

According to the model, we should observe continuous and uninterrupted sequence of the development of every bright object!

This claim should give us the ability to see the uninterrupted development of every luminous object form present position back in time to its first appearance! That means if we are observing expanding universe, we will see every luminous object of the sky like continuous bright line, going away from us - from the present position and time, back in time and distance, towards the time of its first appearance!

The fact that we cannot see such luminous lines of the gradual development of every object is undeniable fact that expansion scenario of Big Bang theory cannot be correct! - (See the image below)

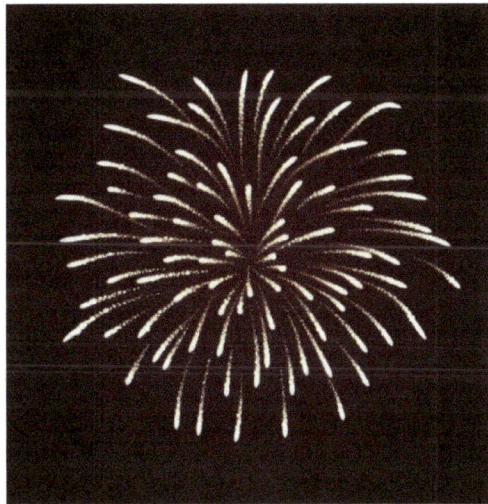

Also, according to the official Model, we should observe real development of the universe, were supposed to be visible the gradual development of each galaxy, each constellation and their similar configuration in all previous stages and distances!

But contrary to the claim we are observing totally different universe on each stage of the alleged 'development' of the universe when in reality we should observe the picture below!

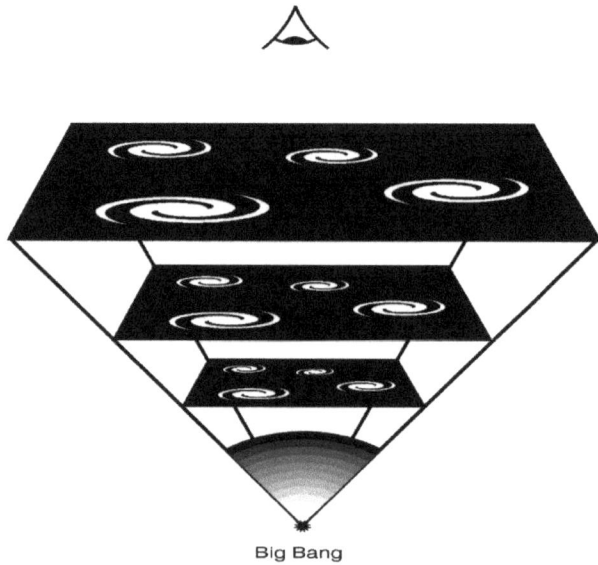

Big Bang

The claim that the universe is homogenous and has been homogeneous in all directions and stages of its development is confirmed with all astronomical observations. The universal structures are evenly spread everywhere from our position to the end of the observable universe! - But aren't you 'smart guys' is the one, who forget to take into account your own claim for space and universe expansion? – What do I mean? I mean that according to "their" claim, the universe should be denser back in time and distance toward the hypothetical Big Bang!

If the universe is expanding, then we should see the distant regions less expanded (1,000 times more dense structures) than the closest to us parts, but our observations are proving that this is not the case! (See the diagram above)

There will be unmistakably visible condensation of the universal structures toward the edge of the visible universe! The fact that we are observing homogeneous density of universal structures everywhere in space, time and distances **is undeniable proof that the universe never been expanding!**

It has same density and structure all the way back to 13.7 billion years ago!

Anyway, let start from the beginning and consider all the claims carefully:

The universe is 13.7 billion years old and is expanding:

We just found that the claim for the expansion of universe is false! They are claiming that the red-shift of the galaxies' light is proof that the galaxies are reseeding form us. But they are not taking in to account the other two physical interactions, which are producing the red-shift of the passing light – Gravity, and electromagnetism.

Gravitational red-shift is a proven fact! - In addition to it, I have to add the wrong claim that the photons are mass-less. The formula $E=MC^2$ proves that any form of energy is actual mass! And any mass inevitably will interact with the gravitational field when it is passing through! All photons are possessing energy, (which is mass)! And this mass is interacting with the gravitational field of the universe when passing through it! The photons interaction with the gravitational field of the universe is producing light red-shift!

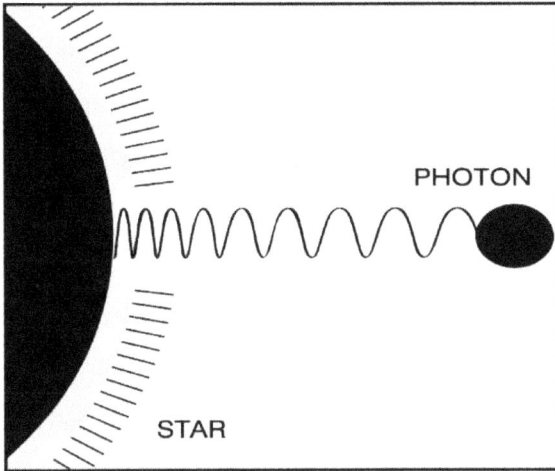

Gravitational red-shift

The physical description of light is: - 'that the light is electromagnetic waves.' Even in the school book of physic, you can find that magnetism, electricity, and electromagnetic waves are components of electromagnetism. The light spectrum is a very wide range of electromagnetic waves, and including radio waves, micro waves, infrared light, visible light, ultraviolet, X rays, and Gamma rays. We are able to see only a very narrow band of the light spectrum but have to know that there is no particular physical difference between the mentioned above parts of the light spectrum except their energy level. We have to know the fact that when the light loses energy, it transfers into the next (lower) band of light spectrum. – For example, if an emitted Gamma ray is continuously losing energy, it gradually will transfer in to - X rays, after into ultraviolet, after into visible light, after into infrared, after into microwaves, and into radio waves. Every educated person knows that the earth's magnetic shield is deflecting most of the sun wind and also deflecting part of the harmful gamma-ray radiation. - This is solid evidence that the light also will be affected by any electromagnetic field where it is passing! The magnetic and gravitational field of the universe is interacting with the passing light, reducing its energy gradually, and producing the observable red-shift, - which is no any indication, that the galaxies are reseeding from us! And this fact is deliberately hidden from the public!

The 'Luminosity' of the universe is undeniable observational proof that the light energy is absorbed on significant distance! We know that in every point of the sky is a star! We don't need scientists to see the dark night sky! - And this is undeniable observational proof that the light is absorbed proportionally by the distance it travels! - And exactly this fact is twisted and presented as proof for hypothetical expansion of the universe as 'light red shift'!

If the light is not absorbed by a few factors, the sky will be incredible bright! The intensity of the coming light will burn everything, and the life anywhere in the universe will not be possible!

20 years later - 1947 in a letter to the Astronomical Society Edwin Hubble state:

'It seems likely that red-shift may not be due to an expanding Universe, and much of the speculation on the structure of the universe may require re-examination' – This time 'they' just chose to ignore him!

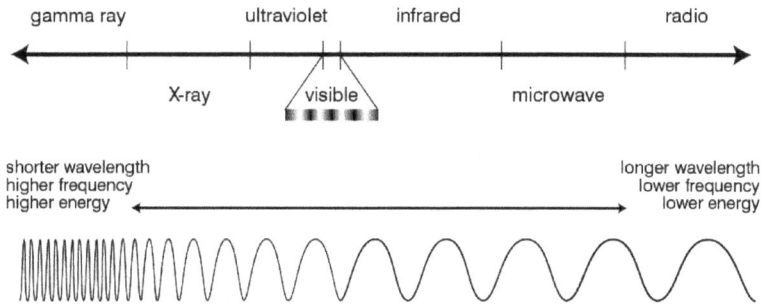

Diagram of the electromagnetic spectrum (the visible light is just small part)

The model claims: that when we look further away, we see the timeframe of past events (or the history) and the last thing we can see is the glow of the Big Bang - (the CMB). Now, we have to take in account the time for the expansion of the universe which supposed to be 13.7 billion years, and the time for the light from the distant position of CMB to come back to us is another 13.7 billion years! The sum of the necessary time for those two events is minimum of 27.4 billion years!

The model is ignoring the necessity of time for expansion and is wrong by 'only' with 13.7 billion years! - This is a well-covered deception, and is not a mistake! - That's why the model is always presented in reverse sequence, so questions won't be asked about the necessity of time needed for the CMB first to distance itself from us on 13.7 billion years before sending its light toward us!

(The universe needs to expand first to its present size and then to send light back towards us).

The universe cannot send light from 13.7 by away before expand and reach this distance! The other problem with this claim is that for the CMB to be at a distance of 13.7 billion light-years (in 13.7 billion years' time existence of the universe), the CMB has to travel (or expand and distance away from us) exactly at the speed of light! - Which is not allowed by the laws of physics and also

CMB will be invisible for us because object retreating with speed of light is invisible for the observer!

The third problem with this claim is: that mainstream scientists claim that the present position of the CMB is 46 billion years away, and is behind the edge of the 'observable' universe! Very well! - These claims are absurd, conflicting and violating the logic and the laws of physics in the extreme, because to be at this distance, the CMB must travel three times faster than the speed of light and have to be in two places at once – at 13.7 by and on 46by away simultaneously, because 13.7 billion years supposed to be also the total age of the universe, and there is no further available time for the CMB to travel from 13.7 billion years distance to 46 billion light years away, which is 32 b.years negative time!

Also, the visible CMB (or Big Bang) supposed to be event, when and where the universe comes in existence! Big Bang (CMB) supposes to be zero time! How do they claim that the present position of CMB is behind the observable CMB, which supposed to be zero time? They are not realizing the sequences of events **that they are going in negative time! - Before, Big Bang!** (See the NASA image of Big Bang!) They actually are adding another 32 billion light years of negative time! - This is an incredible pile of conflicting claims, supernatural events, total stupidity, or limitless arrogance! It is a total mess and has nothing to do with science, healthy logic, and facts!

The fourth absurd claim of this point is: that 'they' claim that we are seeing the present temperature of the CMB. As I stated above, this is not possible, because the light of the visible CMB is emitted 13.7 billion years ago!

The theory says that the universe comes out of nothing! - This is not a scientific statement! And any self-respecting scientist would stop here in consideration of this theory because it is a well-known fact that only nothing comes from nothing! - This is a mystical explanation and has nothing to do with science!

The laws of physics do not allow the matter to be compressed indefinitely. The assumption of the theory that the universe came from an indefinitely small point (singularity) has no physical evidence to support this. This mystical claim is violating the law of physics, contradict the experimental data and the Planck constant – for the smallest distance allowed between particles $(10^{-35}m)$.

The model claims that the universe came from a singularity in the form of superhot energy - (it is not specified what kind of energy) and then turned into matter. For some unknown reason, 50% of all required matter - (the antimatter) is missing from the model! - Scientific experiments consistently disprove this possibility!

The most obvious flaw of Big Bang theory is the mystical assumption that the law of physics and the plan for the structure of matter and universe can come spontaneously out of nothing and form in this superior organize form in a matter of 10^{-38} of a second! Any person with self-respect should distance from such claim because explosion cannot produce an order of such magnitude! The Big Bang Theory considers the visible universe only. And we will also stick to this scenario and will consider only the visible universe:

From our perspective, we are stationary, and everything is supposed to going away from us. That means we observe a sphere with radius 13.7 billion light years and we are in the center of this sphere. - Our observation point always has been in the center of this sphere, even in the time when we didn't yet exist! The starting point for the expansion of the visible universe is the same as our observation point! - (The center of the sphere)

'The Big Bang exploded (or expanded), and the expansion in the first second exceeded the speed of light.' This claim is also a brutal violation of the law of physics because nothing can exceed the speed of light! The point of validation for this claim is not scientific, but is based on the assumption for 'absolute' and 'unquestionable' 'correctness' of the Big Bang Theory! But the irony is that the biggest assumed proof for the "correctness" of the Big Bang is the visible CMB!

But hang on! Here is the biggest proof of the impossibility of BB theory!

The point of origin and direction of the visible CMB actually disproves the validity of the theory with absolute certainty, because from the diagram below is obvious, that the proposed by the BB model CMB is going away from us and have nothing to do with the observable (the real CMB) which coming from the opposite direction toward us! It is also explained that if the visible CMB has been emitted from the distance of 13.7 by. This is undeniable proof that 13.7by ago the Universe has been the same size as now!

(See the diagram below)

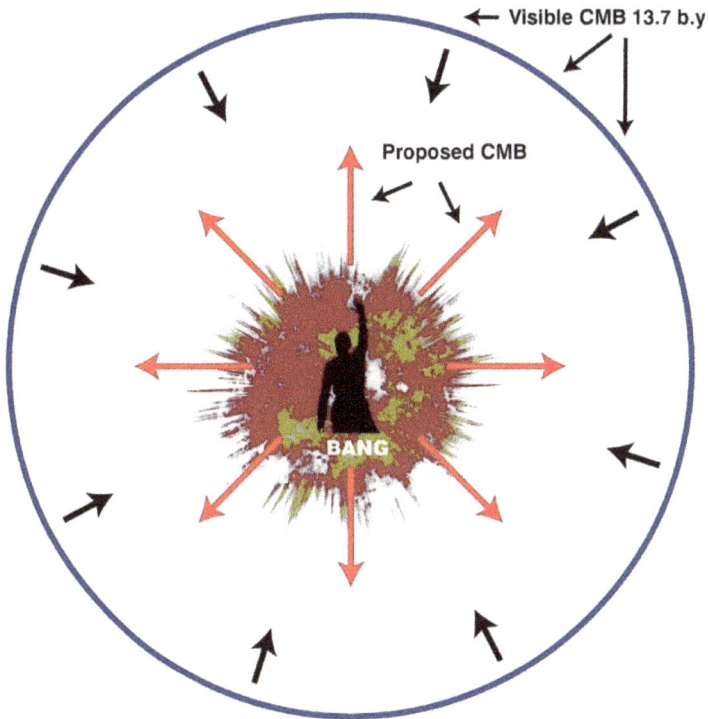

Visible CMB 13.7 b.y

Proposed CMB

BANG

The BB theory states:

In the beginning, the expanding sphere was very hot, and the matter was a hot mixture of elementary particles. After 380,000 years, the hot gases cooled down to 3,000–5,000K. In this condition, the hot mixture started forming atoms, and finally, the light was able to pass (or to be emitted)– the CMB. According to the theory, this is the time – 380,000years – when the presently visible CMB has been emitted. The size of the universe at this time (according to the modelists) has been approximately 90 million light years. We, (or our observation point) at this time, have been in the centre of this sphere (same as now). All the emitted light of the CMB in the first 90 million years must have reached our observation point, crossed it, and continued further in space away from us, never to be seen again! - This is the only time when it has been possible (according to the model scenario) from our observation point to see the emission of Big Bang – (CMB) It is immediately after the emission! But the problem is that we haven't been there to see it! And it is an impossible, absurd claim that we are seeing this emission now!

The visible CMB started its journey 13.7by ago from opposite direction and distance of 13.7 billion light years away in the time when the BB theory claim that the universe has been smallest than an atom and just beginning to expand (see the diagram above). – And this proving the size of universe 13.7by ago!

The model states that the first stars were formed shortly after CMB emission, in about another 400m. years. Those stars were very big, with a very short life in order to explode and produce the heavy elements for the second-generation stars containing heavy elements. – The fact is that heavy element we observe all the way back, even in the furthest visible galaxies 12.4 billion years away, and the presence of the primordial state and such huge first-generation stars is missing from the picture! Why? - The problem with this claim is that there must been trillions and trillions of such big stars, and the rate of continuous supernova explosions in this short time must have been like the New Year fireworks. Such a bright and extremely luminous event with our sophisticated telescopes we cannot miss! Yet we have not observed any such events and primordial state of the universe with our telescopes! And this is solid evidence that this claim is absolutely wrong!

The next problem with this claim is that the big stars, when exploding like supernovas are supposed to always produce black holes. - Such an enormous number of black holes in the primordial time inevitably would have cannibalized all the existing matter of the universe. And there would be no universe, no stars, and no us! - Just black holes and dark vacuums with no thermal energy. The fact that we have not been observing such a grim scenario is further proof that this claim is impossible and cannot be correct. Also, at this point, I would like to add the fact that we observe a fully mature universe 12.4 billion years away, via Hubble deep field photos. But we know that 1 billion years is absolutely not enough time to form a fully developed universe full of second-generation stars! - To validate this superfast unnatural hypothetical development of the early universe, 'they' have invented the mystery of dark matter. - Well done! Another supernatural element with no proof!

The next claim of the model is: that the universe always has been homogeneous and uniform in all the time of its existence: - The model claims that the hot glow of the Big Bang as glowing shell has continuously retreated and is presently on the outer rim of 13.7 billion years away, and it is actually the observable CMB. This claim is ruled out by the validity of the claim - that the universe is, and has always been homogeneous because such a shell has to be made of the material of the primordial matter - gas and dust, - which is absolutely different substance from stars and galaxies.

The next problem with these shells is that, if matter stays in the form of shells, the stars will have no material to form, but if the stars and galaxies are formed, there will be no matter left for such shells! - These are conflicting claims and proven wrong from our observations.

The next nonsense of the claim for retreating glowing shells is proven wrong also with the first statement, that the further we look, we see the timeframe of past events. Fortunately, our tools are sophisticated enough to be able to see everything, except the glowing shells which are missing from the picture. Our observations are concrete solid proof evidence that such shells have never existed. It is a false, unobserved, and misleading claim!

The next impossible scenario with this glowing shell, or CMB, is The assumed "fact" that such emissions happened in the early timeframe of 380,000 years after the alleged Big Bang. At the time of the emission, we are supposed to be in the center of the glowing ball. At this time there were no stars and galaxies. The emitted light had to come to our observation point first, before the formation of the stars and galaxies. But how, in the absolute opposite of the sequence of the theory, did the galaxies and stars get between the glowing shell of CMB and us? The stars were formed second, and therefore their light must come second, to follow the glowing CMB! - That's why we must observe the model in normal sequence; - to be able to see the real sequence of the events! - This is an absolutely impossible proposed scenario, a wrong claim with misplaced positions of the universal structures. It is a deliberate deception and an insult for an intelligent person!

Let us continue with the development of the universe of the Big Bang theory: The model claims that the rate of expansion of the universe is much slower that the speed of light. Even 'they' state that in the first 7 billion years the expansion of the universe has been slowing down, and then the expansion started slowly to accelerate because there is 'mysterious' substance of 'dark energy.' So... they are claiming actually that the universe is expanding much slower that the speed of light! - OK? - No, this is not OK! This is the next false claim! The model and all the observations 'undeniably prove' that the size of the observable universe is 13.7 billion light-years in each direction! That means that the edge of the observable universe - CMB is at a distance from us of 13.7 billion light years. So...the expansion, according to the observational facts, must be exactly with the speed of light –(13.7bly). And then, from this point, the emitted light from the CMB need to travel back to us another 13.7 billion years! The sum is 27.4 billion years! How come? The slow rate of the claimed expansion of the universe cannot match nor explain the 13.7 billion light years size and assumed the age of the visible universe! This is a total mess: conflicting claims, missing billions, missing glowing shells, missing stars behind CMB, missing primordial universe missing first generation stars, and there is a mature universe immediately next to observable CMB, and on the top of this mess, all the numbers don't add up!

The model presents us with another nice mystical surprise! - It claims that the actual present position of the glowing shell of the CMB is 46 billion light-years away! Fantastic, brilliant! The problem is that there are a few impossible assumptions for this to be correct:

If the visible CMB is on 13.7 billion years distance, and the age of the universe is exactly 13.7 billion years, how and when did the glowing CMB manage to jump from 13.7billion to 46 billion? When did this happen? - In no time? Such an impossible thing we call miracles! But even if we accept miracles, this glowing shell have to travel three times faster than light to cross a distance of 46 billion light years in a period of only 13.7 billion years! They claim that the 'Space is expanding'!–Is there any evidence for such claim? Or this is just the next invented fantasy? (Remember that space is flat and the amount of universal components space, time and matter is constant and related?) – That proves that space cannot expand! The light will not travel in strait line if space is expanding, and we will observe different light speed caused by the expanding space! But even if we accept this, it will not make any difference! - We will not be able to see any object that retreats from us with such speed, which they claim to be three times greater than the speed of light! - These claims are conflicting, violating the laws of physics and are deliberate arrogant deception! Such claims are a brutal insult to all intelligent people!

Probably it may be difficult for you to follow the logic of the argument and you cannot fully understand it. Don't worry – there is no logic! They also cannot understand what they are talking about!

I cannot understand how such supposedly very intelligent people can put in one theory so much conflicting claims and stupid assumptions!

They even claim that they can hear the sound of the Big Bang. –Amazing, isn't it?

The correct description of this model is the famous sentence in physics– That "**it is not even wrong**"!

Let's consider now what can glow and what cannot glow: - (The claim for glowing retreating shells)

For something to glow, it needs thermal capacity; that means it needs matter. Vacuums have no thermal capacity and cannot glow! How long actually does the heated solid matter glow before losing its energy? (The formula is SI=J/K.) I will give you an example: - as soon as you switch off a light globe, its stops glowing in an instant, because the heated element loses heat energy very rapidly. Another example is when you heat a brick or a big chunk of steel to 1000 degrees. The next day, when you check the temperature, there will be no difference between the object and the surrounding air! - One day is enough for a solid object to lose all its energy of 1000degrees! Solid objects preserve thermal energy better than gas. And also, the air acts like insulation! The deep vacuum is acting as a vacuum cleaner for thermal energy! The hottest thing we have produced was a thermo-nuclear explosion, where the temperature is above 10 million degrees, but this fireball lasted no more than an hour! A supernova is a small version of the Big Bang and is visible only for several days or weeks; after only a year its temperature is insignificant! How, they claim that the gas cloud of the CMB was only 3.800K 14 billion years ago, yet it is still glowing? When even molten planets manage to freeze in much, much shorter time?

Let's consider the property of the vacuum in our universe:

How much thermal capacity does the vacuum have? - Zero, nothing, zilch! Our universe has an average density of 5 atoms per cubic meter, and these five atoms are mostly concentrated in the galaxies. So, a realistic density figure will be a few atoms per cubic kilometer of space. I am sorry, but emptiness or vacuum with such density cannot preserve any energy more than a minute!

The surface of celestial bodies varies between 300 and 500 degrees day and night-(Mercury). Even such massive objects like planets cannot preserve thermal energy for longer! How is it possible for an extreme vacuum to glow continuously for 14 billion years? Currently, we have observed unchanging temperatures of the observable CMB for 50 years. Any physical object (even solid matter) will lose a lot of thermal energy in this period, it will cool down dramatically, but we do not observe any change in the CMB temperature. Why? Even if you spread all the matter of the universe evenly across the space of the universe, the density will be five atoms per cubic meter and the universe will lose its thermal energy probably in a day (according to the Dulong-Petit Law). It is time to realize that only the energy source of distant stars and galaxies can provide such continuous and constant glow! -**Yes! The visible CMB is the light of the distant galaxies of our Universe - beyond the proclaimed end of the Universe at 13.7 billion light-years away!** It is the light of the endless and vast Universe that sends us the gentle, inviting glow, reminding us that is time for us to become more honest and start to study and learn the real secrets and puzzles of this amazing world!

We also observe a very strange phenomenon. In one part of the sky, 800 galaxy clusters stream in one direction, obviously attracted by some huge structure, which is not in our visible range and are presumably outside of the observable universe- (the Dark Flow). This clearly reveals that there are some extremely massive structures beyond the observable horizon.

If there obviously are such huge structures behind the observable horizon, why are the leading scientists refusing to look further than the end of the universe assumed by them? Is this an accident or orchestrated deception? Actually, we have evidence, that the universe is much, much bigger than the assumed 13.7bly because the observation data from ultra-long radio wave telescope in Tasmania show that light coming from every direction far behind the assumed 13.7 boundaries of the universe!

We are observing microwave radiation coming from deep space. But this CMB which we observe is behind the galaxies, coming from the opposite direction of the proposed radiation of BB and cannot be associated with any Big Bang! - This radiation is the glow of the distant galaxies of our enormous universe were the distance, the gravitational and electromagnetic field of the universe is depleting the energy of the emitted light from these distant galaxies, and their light appears in microwave spectrum! And that's why the temperature of the observable CMB is not changing for 50 years! And this proves that its source cannot be glowing matter, but is the constant glow of distant galaxies red-shifted to 2,725K. Any matter inevitably will cool down in 50 years! Remember even the surface of planets cooling down during the night a few hundred degrees! This is fact proof that the universe is much larger than what the crooks try to convince us, is not expanding, but sending us the continuous glow as an invitation to study and explore this vast and wonderful universe! All the calculations for retreating galaxies were done before the alleged "discovery" of Dark Matter and Dark Energy and have not been changed. If the leading scientists really believed in the existence of Dark Matter and Dark Energy, they would make the necessary adjustments in their calculations, but the fact that they haven't done this is obvious proof that they knew and had been hiding the fact that Dark Matter and Dark Energy do not exist! - Only this simple fact can tell you the real story behind this absurd model.

If the 'dark stuff' is taken into account, the gravitational red-shift will be 24 times stronger than the original calculations. Assumptions based on 2,400% inaccuracy are too big to be tolerated and assumed to be correct!

If we leave aside the weird theories, skepticism and supernatural explanations, the logical answer is simple: The Universe is much bigger, much older and there are massive structures behind the observable horizon. The visible CMB is the furthest part of our universe visible in the microwave range because the distance is depleting the light energy. There is nothing mysterious in this, just simple logic in following the evidence.

So, in the end, what actually our conclusion is?

When we have considered the claims and development of Big Bang model of the universe in normal sequence, and with correct timing and place of events and without the necessity of violating the laws of physics, the logical and obvious conclusion is that:

All the claims of the Big Bang Theory are incorrect, mystical, supernatural, conflicting and violate the logic and laws of physics with misplaced positions, incorrect numbers, hiding data, and make mystical claims. There is no single correct point or assumption in this theory! It is more than obvious that the Big Bang Theory is neither correct nor scientific. It cannot be classified even as theory! It is of no value to science, and even is not appropriate to be accepted by any religion, because it is false! The time is up for this absolute mess and deception created a century ago to go into oblivion and open the road for real science and knowledge!

We cannot build our knowledge of misassumptions, deceptions, and wrong models. The model of Big Bang is built on properties that violate logic and laws of physics, and are based on weird mystical and supernatural assumptions, misplaced points of directions and observations. Such kind of science cannot serve us well in search for understanding of the world.

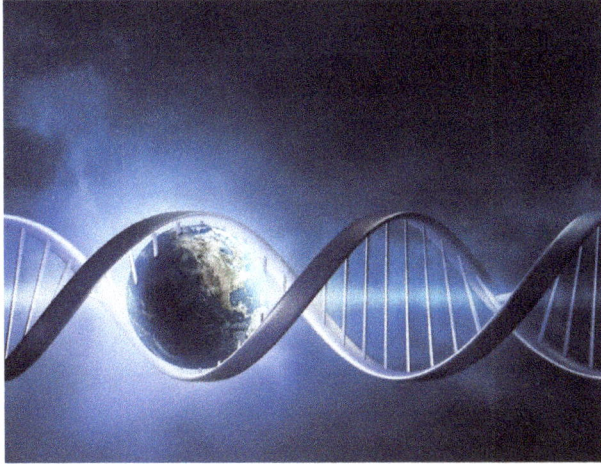

The origin of life is probably the most important question for our understanding of the world and ourselves. Because of its importance, the answer is subject to many different explanations, but like everything important, in the explanations are always inserted some hidden agendas. In general, if mystical elements are not inserted, some facts are hidden, others are exaggerated, or the explanations suffer a lack of competency and lead to totally incorrect conclusions. It is important to know the origin of life, but it is also good to know the details and the origin of the theory that explains the origin of life. This will give an indication for the possible inserted agenda serving the authorities, policies, or interests. The elite are not far behind exploiting this vital question for its own purpose - to create confusion.

It is no accident that the most popular explanation and theory of nature, promoted by the media as the most credibly and "scientific," comes from a person who "mysteriously" was awarded the rare privilege of having a privately funded scientific expedition by ship around the world. Probably you have already guessed who this person is – Charles Darwin.

It is a very revealing coincidence that later Niels Bohr, the author of the other fundamental doctrine for our understanding of the world, the 'Copenhagen Interpretation,' was also privately funded and had a privately funded institute. Nobody else from the scientific community in the past has had the luxury of such generous funding. The outcome of this generosity was the establishment of the two fundamentally incorrect (but very comfortable for the elite) theories for our understanding of the world. It is a striking coincidence, also, that both - the Big Bang and evolution theories use the same method to prove the origin of the world and origin of life, working backward and ignoring fundamental principles and evidence, which prove their incorrect and false basic assumptions.

I will give another example of how much money and effort the elite allocate for the tools of public deceptions, propaganda, and chaos. Not many people know that the establishment of communism in Russia served 'their' philosophy and agenda perfectly, because the communist system was the most effective tool for the destruction of the Orthodox Church, for the propaganda of Darwin's theory of evolution, and as an example of a badly implemented system of social equality. Everybody believes that the Bolsheviks instigated the revolution in Russia, but the reality is that this was a well-funded and organized coup of the Western elite. History tells us that Lenin and his entourage were transported by train through two front lines during the First World War from London and Switzerland and on to St Petersburg where, instead of being arrested, they were well funded. The bank record shows that 34 million marks (equal to billions in today's currency) were sent to him from German banks through Finland. How much England contributed is unclear, but a large part of the Bolsheviks army has been armed with English rifles. That's how Lenin was able to organize and supply his army to topple the government. The cruelty and repression of the Bolsheviks became a propaganda tool for the Western elite to control and scare their own people. This is a good example of how well organized the elite are, and even the war could not break their cooperation and collective actions.

It is a similar story with the "freedom fighter" Yasser Arafat – his people were dying in poverty, but he ended up with 4 billion dollars in his Swiss bank account.

And now... to come back to our main subject, the origin of life. We have to consider the origin of life on Earth with an open mind and without any prejudice, because the origin of life has only two possible sources: spontaneous or introduced. Either of these two possibilities needs solid evidence to be able to be accepted as the most credible explanation for the origin of the life on planet Earth.

We have to consider what the details of the main explanations for this mystery are. The first of them is the classical explanation of all religions – that God creates life. I am not able to make any considerations for or against this, because religious explanations are based on faith, not on facts. This explanation belongs to the 'introduced life' category. The second explanation that we have to consider is the main scientific theory for the origin of life, where Darwin's Theory of Evolution is the central fundamental assumption.

It is a widely held opinion that Darwin is the author of the idea of the adaptation and evolution of species, but this is wrong. At the time of the publications of his papers, the scientific community had a good understanding and enough evidence for the natural process of evolution. For thousands of years, people had bred different domesticated animals and plants using the technique of selection. Claims that Darwin discovered this is absolutely incorrect. The fact is that the young Charles Darwin (22y), who study art and haven't been even biologist has been privately funded and send on HMAS Beagle with a "special" mission. On his return has been promoted to celebrity status by the media, and given Treasury grant of about $150,000 (present value)! - These facts are reviling who stands behind his "story" because the concept of Darwin theory is "the survival of the fittest," which actually postulating that violence is the cause for evolution and progress! This doctrine has been adopted by the elite and brings us endless wars, sufferings, and inequality. After 23 years in 1859, Charles Darwin finally has delivered what he has been paid to deliver - he published the book, 'On the Origin of Species.' The main concept of his theory is the assumption that life on Earth has evolved by chance from a mixture of organic compounds in the primordial pool. The assumption is that all living organisms have common ancestors from the past. The evolution of the living organisms is done by natural selection, to preserve and accumulate minor advantages in mutations of organisms and pass them to the next generations. This preservation of minor advantages enables organisms to compete and adapt better to the natural environment. Some use the term 'survival of the fittest', which is short but not the best description of the process of evolution.

Darwin realised that evolution is a slow and gradual process because complex organisms are very sophisticated, and all body systems must evolve in parallel for the organism to have a chance to function properly and to survive. In the main, the specimens and his observations of nature, which Darwin made during his expedition on the Beagle, provided him some evidence needed for his theory. At this time, the science of genetics was unknown, the paleontological record was not very good nor consistent, and the methods determining the age of the specimens were not reliable. In these conditions, I have to admit that Darwin has contributed to the idea and our understanding for the evolution of species, but on the other side, he sent our research for the origin of life down a dead-end road. The ability of living organisms to evolve and to adapt to changing environmental conditions is one thing, but this ability in no circumstances can be used to explain the origin of life. And yet Darwin did exactly this! From the time when Darwin published his theory to the present day, scientific methods, instruments, and techniques for classifications and dating of specimens have improved dramatically; the accumulated evidence provides us with a very good picture for the development of life on Earth. There are millions of well-documented examples of living organisms from all periods of Earth's biological history. There are millions of specimens, but there is a problem, and this problem is significant! The biological record and the new understanding of organisms' true genetics do not support the Darwinian theory for the origin of life, because the enormous complexity even of a single cell cannot be explained by the spontaneous organizing of chemical compounds into such a highly sophisticated biological system which even the most primitive single cell possesses. The functions and DNA of the most primitive cells surpassing the complexity and sophistication of all software programs on Earth. Darwin proposal is that such sophisticated and logically constructed software of biological life can arise spontaneously from randomly mixing of chemicals in the primordial pool of water. This is an absurd proposition because DNA mutation actually is damaging the functional order of the existing DNA! There is also a huge problem with the discontinuity in the developmental line of the specimens through the past geological periods and the sudden and (stepped evolution) - inexplicable appearance of new and more sophisticated complex organisms without any sign of predecessors. According to Darwin, we should observe a continuous, uninterrupted development line of all spicy, but we are observing sudden appearance and disappearance of fully developed spices. The claim that the evidences have not been found yet is not evidence for the correctness of Darwin's theory for the origin of life. We have a very rich paleontological record from all geological periods, a very good record even for single-cell organisms far back to 4.1 billion years ago, but the common ancestors of the complex cell organisms are just not there. This is a very significant fact. We witness bacteria, unchanged from the beginning to the present day, but we never witness any evidence of slow and gradual developments of complex-cell organisms, where different colonies of bacteria have to transform themselves into specialized organs with specific body functions.

Here is a brief history of the development of life on Earth through the paleontological eras. For easy classification, I will use the major periods:

- Pre-Cambrian (the boring billions): 4.5billionyears to 600 million year ago, where only single cells have been found
- Cambrian: - the explosion of life 600 million years
- Ordovician: 505 million years to 438 million years, where the earliest land plants have been found
- Triassic: 245 million years to 208 million years, where the first mammals have been found
- Jurassic 208 million years to 146 million years, where dinosaurs and the first birds have been found
- Cretaceous: 146 million years to 65 million years. ended with mass extinction
- Cenozoic (the last three periods): 65 million years to present, where mammals and humans exist.

As usual, there is confusion within the present scientific community about every detail of the paleontological and geological history of the Earth, but I will try to give you a brief sequence of the major events with a full understanding of the chaos in the scientific data:

The earth's formation is assumed to be about 4.5 billion years ago. In a very short time, almost immediately after the earth's formation, the earliest biological life on Earth around 4.1 billion years ago began, and evidence for this has been found in sedimentary rock in Western Australia. Actual bacteria fossils have been found to be 3.8 billion years old. Between 4 billion years and 3.8 billion years ago, the Earth suffered a heavy meteorite bombardment. The consequence of this is controversial but, if there had been bacteria, they managed to survive this carnage. The appearance of the first bacteria shows high sophistication and ability to process minerals and sunlight. The earliest oxygen traces are dated to 3.5 billion years ago, produced by prokaryotic and eukaryotic bacteria. The source for the atmospheric oxygen is believed to be Ciano-bacteria or blue-green algae from later periods. A colony of ancient oxygen-producing stromatolites is still present in Shark Bay in Western Australia. The oxygen level, in the beginning, was very low because it was continuously absorbed by rocks and sea-dissolved iron and locked in the earth's crust. Finally, in the period of 2.3 billion years ago, the ground was saturated and could not absorb more oxygen. The atmospheric level of oxygen managed to rise and caused the extinction of anaerobic bacteria because oxygen is toxic to them. The rising of oxygen levels is believed to have created the so-called 'Oxygen Catastrophe' because oxygen oxidated the free methane and transferred it to carbon dioxide (CO_2), which is a weaker greenhouse gas than methane. Large parts of CO_2 have been absorbed by minerals to form carbonates. The lack of methane and CO_2 leads to the freezing of the earth in a complete ice shell ('Snowball' Earth). This glaciations period is believed to have lasted 300–400 million years. The palaeontologic record estimates that at least 95% of living cells became extinct. The survival of any form of life in these harsh conditions was incredibly lucky. The combination of low volcanic activity, the absence of methane-producing bacteria and the extremely reflective surface of 'Snowball Earth' looked like they could have sealed the fate of our planet in a thick permanent ice shield. But then something sudden and extraordinary happened. The movement of the Earth's tectonic plates created major volcanic activity, releasing huge amounts of carbon dioxide and snow-blackening ash. This created the conditions for global warming and melting the ice shield. The Earth has been free again!

There are a few hypotheses about how life survived those periods of glaciations. One suggestion is that life survived in the hydrothermal vents; another suggests that narrow open belts of water remained in the equator. But the fact is that for the period from 4.1 billion years ago to 600 million years ago, Earth was transformed, had its atmospheric composition changed and suffered major extinctions, yet life remained in the form of single-cell organisms for 3.4 billion years.

The last glacial period ended a few million years before the Cambrian explosion of life. It is logical that in this glaciations period, we cannot find any multi-cell life forms, because the extreme conditions of glaciations did not provide any food sources or conditions for sophisticated organisms to evolve. Apart from a few controversial Ediacaran fossils, the paleontological record confirms the absence of complex organisms before the Cambrian period. The Cambrian life explosion starts with many (about 600) different complex cell organisms, distinct from each other. The principal difference between single-cell and complex organisms is the enormous amount of informational organization involved in complex life. Such packages of incredibly sophisticated informational structure cannot come suddenly from nowhere. The complex cell organisms cannot evolve in such a short period as the time for the gradual evolution of such sophistication is measured in hundreds of billion years. Opposite to the theory of evolution, the anatomical variety immediately after the initial diversification of the complex cell organisms reaches its maximum. From this point, we are observing eliminating, not an expansion of the basic anatomical design of species.

We know that packages of logical information usually are the product of an intelligent mind. And this is the puzzle! Where from this package come from? The incredibly short time for the appearance of such rich biodiversity in the natural system makes the Darwinian explanation for the gradual evolving of organisms absolutely impossible. The complexity and sophistication of the newly appeared organisms have been incredible. The single-cell organisms remained single-celled for 3.4 billion years without any sign of advancement towards complex cell sophistication. The slowly evolving and accumulating advantages in single sell organisms have not been observed at all. Instead, we observe a sudden explosion of completely new forms of life organisms, where the genetic information for such sophistication is assumed to need longer to evolve, much longer than the assumed by the Big Bang theory age of the universe, which is 13.7 billion years. This fact leads to much-heated debate and countless theories, where the correct point being only the fact that life on the Earth has appeared and evolved into its present form. But this fact is not proving anything! The problem there is that all those theories suffer a lack of evidence, logic, and scientific and evidential credibility.

I will explain the absurdity of the situation with a simple comparison. Suppose you found a typewriter in the jungle with a piece of paper on with 'Hello, my friend' written on it. Your two companions start convincing you of two possibilities for the origin of this typed phrase, one insisting that this is the work of man, the other insisting that monkeys, when playing with the machine, accidentally typed the phrase. There is no evidence to prove or disprove either of these claims, which claim will you accept, when you calculate that the possibility of the monkey accidentally typing the phrase is 1in 36 trillion! It is the same case for the probability of life emerging accidentally, except that the odds are many trillions of times bigger than the example with the monkey and typewriter.

Another example: in a box, you have 12 books. How many times do you have to put them randomly on the shelf for them to be in alphabetical order? With only a combination of 12 sequences, the answer is 5 million times. We are now faced with the sudden emergence of cells whose complexity of DNA and sequences is on unprecedented magnitude of billions. How long does nature have to play at random roulette to have the chance of creating even the simplest correct information for the simplest cell or complex cell organism?

The chance for this to happen accidentally is trillions and trillions and trillion times more impossible than the accidental monkey typewriting. If we cannot accept the "accidental" origin of the simple typed phrase in the jungle, how we can accept the possibility of something far more impossible? Yet this is exactly what the elite want us to believe that such an impossible scenario has to be the most credible explanation for the origin of life. To be able to have the real picture, we have to find how complex the single cell is and what exactly the real story behind the mystery of the origin of life is. We have to go deeper in the structure of the living cells, their functions and the complexity of DNA, how they function, how they evolve and what is the information role in the structure of living organisms.

To understand how life has arisen, we have to have a real understanding of the universe where we are living. The theory of the Big Bang is wrong; the universe is most likely endless and eternal, giving us a different prospect and possibility for our understanding of life's origin because the assumed 13.7 billion years are absolutely not enough for such sophisticated life to have arisen slowly and gradually. But if the universe is much, much older, this makes a big difference for the possible source of life. The presence of organic material everywhere in the universe makes the possibility for life absolutely certain. The eternity of the universe allows the Theory of Evolution a bit more time, but even this still does not provide it with enough credibility for an answer. Only when we have a clear picture for the correctness of the facts will we be able to have a credible opinion of what is possible, what is not possible and what is pure speculation. To know the origin of life is important because we will know who we are, where we come from, and what is the purpose and future of our existence.

For all the periods of life on Earth, there exist only two types of bacteria classified by their structure: prokaryote and eukaryote. Prokaryotes are the first bacteria to appear on Earth. They are simpler and have no nucleus. Prokaryotes remain single-celled while evolving in many different ways. Despite these variations, the prokaryote structure has remained the same for the last 4 billion years. The eukaryotes are a more sophisticated single cell, appearing approximately1.2 billion years ago. They are about 10–15,000 times larger than prokaryotes. Eukaryotes have a nucleus, and one can say that they are identical to the cells of all plants and animals. We have to understand that the basic provider and life support for the eco-system of Earth are provided by bacterial life. Bacteria have spread and adapted to live in each part and environment of our planet, from the air to the rocks kilometers deep in the earth's crust and even in the acidic juice of our stomach. Each gram of soil has an average of 40 million bacteria. There are about 1 million of them in each gram of fresh water. The number of bacteria in our body exceeds the number of our own cells. Their biodiversity is enormous. Currently, less than half of them have been studied. But there is a realization that even the simplest bacteria are an amazing complex of sophisticated biological and information systems and the attempt to explain this sophistication with an accidental assembly of organic matter is absolutely unacceptable. Currently, our scientific laboratories are not able to produce even single element of the most primitive bacteria! How dirty pool of water can produce such sophisticated structure as living cell?

Let's see how sophisticated the structure and functions of some of the systems of a single cell are. Everybody thinks it has just a simple, flexible membrane. But the cell membrane has many roles in the structure and functions of a cell: to be strong, to be flexible, to be water-based but water-resistant, to provide propulsion, to allow transfer in and out only necessary substances, to sense those substances, to sense the environment for light and to make contact and communication with its own kind. To be able to do this impossibly complicated task, the membrane also must have an information system, a program for its task, and tools to read the information on which substances to allow in or out. It also must have a sophisticated 'control gate' for the substances to pass in and out. If the membrane cannot meet all those requirements, the cell simply cannot exist. And we witness such sophistication in the appearance of the first cell. Where does the information program for this complicated task of the membrane come from? Definitely, evolution is not the answer!

Information reading system and a production assembly line of the living cells is something far more complicated even than DNA by self! How can you make soup of chemical substances to read information and then to obey the informational order and start producing the necessary complex proteins? The sophistication of such system is surpassing any imaginations! I will give a few examples, which will describe the complexity of this cell functions: Imagine instead of DNA I have memory stick where is written a file with Opera, file with diesel engine, and file with house plan. Just to read this information I will need a computer, and if somebody suggests that if I shook a bag full with electronic parts, they can spontaneously arrange themselves and form computer for me to read the files, this will be the most stupid suggestion I ever heard. On top of this, to play the music, I will need symphony orchestra, musical instruments, opera singers, music hall, and musicians, with decades training! Even if I get a bunch of university professors, they cannot play the music! To build the diesel engine, I will need engineers, factory full of machinery, workers, bearings, special materials, and electricity. - From those examples becoming obvious, that the building process and realization of the biological information – (DNA) is much more complicated than the entire package of logical information! In reality, random mutation is not producing more sophisticated forms of life, but greatly create disorder in the well organised system!

Let's see how the cell assembly line works. - The protein assembly tools of the cells are called ribosomes. The ribosomes are incredibly tiny. A single cell contains up to 13 million of them. They receive specific orders for production of different proteins, read the instructions, decode the structure of the protein into the building blocks of amino acid chains of sequences letter by letter, and then produce the amino acids with speed and at a rate of 10 amino acids per second. Then they assemble the chains of the different amino acids in a precise order to form the required protein. The size and the sophistication of the ribosomes are incredible. They are tiny and are compared to an atom. They work with a precision not achieved even by our computers. They work with a speed unparalleled to anything that has been created by us. They have their own software program for each task, the tool to read this program, sensors for performing their task, energy for self-sustenance and work, and communication systems to receive and read the specific orders. We face again the same question which evolution cannot answer: where does the information code or program for these sophistications come from?

The other puzzle we face is the similarity of the building blocks of all life – the cells. The cells of all plants and animals are virtually the same. The famous biologist Nick Lane has stated: 'If I look and compare the cell of a mushroom and mine, I cannot see any difference.' What determines which will be the mushroom and which will be the biologist? It is obviously the information embedded in the DNA, but where does this information come from? It is more than obvious that a pile of dirt cannot organize itself into a living organism. Evolution has no answer for where this information came from because the simple fact is that the most primitive single cells - amoeba, have more DNA than us! Does that make sense according to the Theory of Evolution? The DNA of one single cell has more information than all the books in the libraries in the world together in one row. Can I tell you one fact? Onions have 5 times more DNA than humans. Have onions ever been five times more advanced than us to have such superior DNA and then reduced its sophistication back to an onion state? - How can evolution explain this? The logical answer is that the package of biological information has come in sophisticated form, and the organisms are using only the necessary for them part!

We must face the fact that all life on Earth is related, and has arisen only once in the last 4 billion years immediately after Earth provides conditions that allowed life. This fact rules out the possibility for spontaneous arising of life because if life can arise spontaneously, it should be able to rise again and again and again, because the present conditions are more favorable for this. Yet we never observe anything like this. When we look for logic in the history of life, two major events look suspiciously like an intervention. The first one is the appearance of life immediately when it was possible, like someone was waiting to give life the best possible start and more time to succeed, and the second occasion was the Cambrian explosion of life where, in the blink of the eye, about 600 different types of complex cell organisms and countless plants forms came from nowhere. I say nowhere, because we have not observed the long period of slowly accumulating evolving advantages, but a multi-lined parallel explosion of sophisticated life forms.

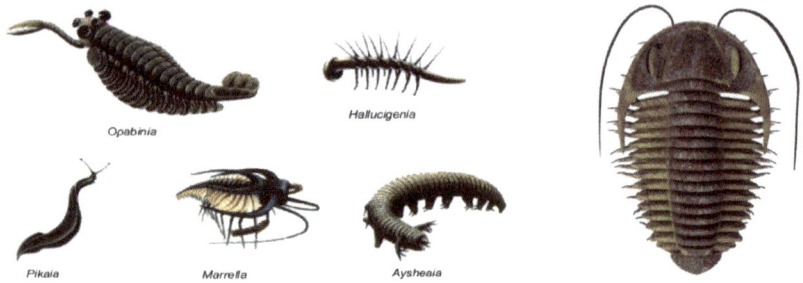

Sophisticated organisms of Cambrian life explosion

It looks very much like when conditions became suitable, someone inserted and spread the information blueprint with the formula for complex life and le evolution to complete the job. The content of this formula of life (or program also revealing the logical vision, because there is an embedded system for continuous sophistication, which we do not observe in single cells; - they remained virtually the same for 4 billion years! This mechanism for sophistication has very distinctive features – a mortal body with an immortal sophisticated information program. This scenario also looks like introduced, because this package of information appears very suddenly, like from nowhere. Our increasing knowledge gives us an understanding of the real structure of living organisms, and this new understanding is that ordinary matter (pile of dirt) cannot create a life because it cannot produce and provide the necessary information! Only the information can create life! The foundation of life is its blueprint, its information, the structural plan, the idea Living organisms can use many different substances or minerals for the body structure, but the information code remains the same! When we consider and analyze all those facts, the officially supported theory for 'accidental, spontaneous' origin of life looks like a very naïve and impossible scenario. Thi theory has been created and supported by the rich elite and serves their philosophy, but not serve the truth.

When we connect the dots of the puzzle, it's getting obvious that the evolution theory not providing a credible answer to the appearance of earth life. The logic of the available data and facts, giving much more credibility to the scenario of introduced life. The fact that the first bacteria have produced huge variety, but their fundamental structure remains unchanged for nearly 4 billion years is telling us that the first bacteria has been deliberately designing not to evolve and have been introduced with a special purpose - to create the conditions for complex life organisms. One cubic inch soil contains more bacteria that the number of the entire human population. Those bacteria are producing the nutrients for all living organisms on earth. The complexity of their chemistry is surpassing even the capability of any laboratory, or chemica plant. If those bacteria evolve, the production of nutrients will stop, and the life on earth will spiral toward the dead end! If the theory of evolution is correct, the primitive bacteria inevitably should evolve in more complex and sophisticated life forms, but looks like they have been designed with special purpose this not to happen! And when we are facing such chain of the logical events in the emerging biological life, we have to use our knowledge to make sense of the facts available to solve the jigsaw puzzle of the life. If we would like to solve this puzzle, we have to abandon the obvious primitivism on which is based the officially promoted idea for the spontaneous origin of life. The emergence of life is closely related to the puzzle of the origin of the universe! The science is completely muted on the origin of the law of physics, and the origin of the quantum information, which is the (DNA) of matter. Similar scenario we faced with the origin of universe where the matter cannot spring spontaneously, because is made from single substance – energy, and the law of physics and the quantum information must exist before the appearance of matter to provide instructions how the energy to be transform in matter!

The complexity of quantum information and law of physics is enormous!

All those facts about the law of physics and the quantum information are accepted as fact, but the establishment does not allow any comment on this vital subject, because exactly this is the vital knowledge for a correct understanding of our world! Behind all those hidden laws and information is lurking the shadow of consciousness! It's getting obvious that consciousness is the possible creative element of the puzzle! It acts as the hard drive of the universe, where the quantum information and the law of physics are stored and operating from!

If we accept the scenario how the universe work, then it will be easy to understand where from the secret informational code of life is coming from! – Is coming from the same origin, where the law of physics and quantum information is coming from! - The universal consciousness!

We are facing the fact that the necessary condition something to be alive are to poses consciousness! And we know that all living creatures, even the simple bacteria possessing consciousness!

We don't have to accept primitive explanations for the origin of life when we have an understanding of the real structure of the universe, its properties, and the way how they are working! This crucial knowledge has been carefully hidden from us because the knowledge is power!

The paleontological record clearly shows that the life on earth going from primitive forms toward sophisticated organisms, but the paleontological record shows us something very significant! The anatomical variety of species reaches its maximum shortly after the appearance of complex cell organisms! The present biodiversity is based on fewer anatomical basic designs, which show elimination, not increasing of the new life forms! And this is the smocking gun! The paleontological record show also sudden and unexplained appearance of new types of sophisticated species without a trace of gradual development! The present discoveries in DNA show that some less advanced organisms have five times more DNA than humans! The onion has never been more advanced than us! –It is obvious that the onion has inherited the sophisticated DNA package and using just part of it!

When we put together all these facts, they are leading to one logical conclusion: The lack of gradual development and transition chain of developing species in the paleontological record leading to the only credible scenario, where the complete package of biological information for future development has been inserted in the early stage of living organisms. Only this can explain the available paleontological record and can explain the sudden appearance of new fully functional and sophisticated species without transitional development! It looks like in the original package of biological information has been included numerous different anatomical designs and left there the nature to choose the most appropriate forms! Yes, the living organisms have to evolve and follow the inherited development plan. They cannot jump over any stage of the pre-programmed functions of body sophistication!

Our increasing knowledge and facts we possess reveal the true nature of the origin of life, the mysterious and magical complexity of logical information and the vision imbedded in the biological structures with a 'computer' like program that have the ability, and destiny, to become more organised, more intelligent and to obtain possession of the ultimate secret and power of the universe –

The intelligent consciousness! But the gift of consciousness and intelligence comes with certain conditions attached – ethics. Greed and cruelty are primitive instincts, and unfortunately, they are the current foundations of our society. They are tools for survival of primitive life forms but is cancer for intelligent society. We have to grow up and understand that consciousness and ethics are the supreme manifestations of the intelligent mind. We won't be allowed to continue further if we do not abandon our primitive instincts and cruelty. - The conditions are not negotiable! Unfortunately, it looks very much like we are doing everything possible not to accept those ethical conditions imbedded in the grand plan of our Universe.

The scary skeleton of *Tyrannosaurus rex*

The chronology of earth geological eras is mainly considered as pre-Cambrian eras – (Proterozoic) where the biological life on earth has been in single sell bacterial form and from the Cambrian period (Phanerozoic) - to present time. The Cambrian period is the most intriguing period because the complex cell organisms suddenly appear with no apparent time for slow and gradual development which evolution theory suggesting. We are having similar confusing stories with the Dinosaurs extinction.

Here are the facts and theories with which we are currently presented and officially accepted for the mysterious disappearance of dinosaurs.

Dinosaurs ruled the earth from 145to 65 million years ago. We have found the indications of their presence in the sediment and layers associated with this period. The last place where the remains of dinosaurs can be found is in the so-called K-T boundary (Cretaceous-Tertiary boundary), represents the time mark of 65 million years ago, where something dramatically happened to wipe out most of life on Earth. After that, we cannot find any remains of these magnificent and mysterious creatures.

There are many theories, but we will consider the four most popular:

One: The earth and its climatic conditions, together with some additional factors, gradually changed the ecological conditions and created the mass extinction.

Two: Volcanic eruption. - In the K-T period layer, there has been found evidence for major volcanic activity in India, the Deccan Traps, which lasted 2 million years.

Three: Impact Theory. In the same period evidence has been found of a major meteorite impact in the present Gulf of Mexico, the Chicxulub crater, 150 km. in diameter.

<u>Four:</u> Tectonic Plate Theory. The seafloor subsided, and this made a big impact on the climate and condition of the sea.

The theories point towards some available facts, but those single facts cannot cover the complexity of the ecosystem, and such simplicity is actually not an answer. This is like trying to find the beauty of a Rembrandt painting with a magnifying glass or a microscope. That's why all of the above theories cannot give a credible answer for the real reason of the dinosaurs' extinction because the extinction was very selective and didn't affect all the species equally at the same time or in the same environment. In reality, not all of the species from this period became extinct. Reptiles, crocodiles, and birds are good examples of this. The second point is that, according to the established view and evidence, the actual extinction period is about 10 million years. Even a recent BBC publication, according to various university studies, has extended this period to 50 million years. I am sorry, but a period of 10 to 50million years is long enough for the climate to recover and species to adapt to new environmental conditions after a catastrophic event. This is why all those claims have no realistic base and explanation for the event of this mass extinction.

For example, wide spread believe is that humans have evolved from primates in a period of only 2 – 3 million years. This has been extensively studied, and if is correct only this fact alone will be enough to prove that 10–50million years is long enough for nature and species to recover from any climate and catastrophic event. The presented theories are considering only a narrow-selected amount of data, and have an ignorant position for other clues and evidence. As a result of this selective approach, they suffer a total lack of credibility. It is obviously that not lack of data, but deliberate misinterpretation and misinformation are the reason for these totally different explanations for the same event. Looks like that the creatures of the past geological period have lived on steroids and their enormous size is not explained by the main official theories! There are examples of enormous size animal life in all environments – air, water, and land! And this fact only can tell us how badly we are interpreting the available facts to construct our most fundamental theories of our life!

Such an obvious clue, like the enormous size and weight of the dinosaurs 60 – 80 tons, even lost skeleton of Amphicoeliasfragillimuswhich indicate weight up to 170 to be overlooked. According to some studies, such enormous weight presently cannot be supported by any bone structure! – This fact also has been overlooked and ignored. This conclusion leads to the necessity of considering the strength of the earth's gravity in the past. Some suggestions are that the earth's gravity at this time was much weaker, and this allowed all animals to be of enormous size compared to present-day examples, such as African elephants which are about up to 8 tons and looks like they have reached the maximum size limit which present-day gravity allows. - And again the dogma inserted in science is not allowing the scientists to study and consider the possible electromagnetic origin of gravity. Variation of gravity could be the real factor behind the disappearance of the massive Dinosaurs and success of the smallest creatures because change the strength of gravity will affect not only the animals but will affect all structure of the eco-system and create cascading effect!

The other ignored scenario for the disappearance of dinosaurs is a potential period of increased galactic electromagnetic activity and radiation, which will shower earth on regular bases with high energy charged particles and gamma rays. Well, known fact is that the extinction has been very selective - the small nocturnal mammals, which in daylight is hiding underground and reptilians have survived, but the animals of the open field have disappeared. It is good evidence in support of this scenario.

We possess a huge amount of data and evidence that has to be carefully considered without bias to have a real picture of the period: fossil records, plant records, climate, fauna, geological and atmospheric data. Yet, despite millions of dinosaur skeletons, bones, eggs, footprints (countless clear evidence), we are presented with incomplete, confusing, and conflicting theories for the extinction of dinosaurs. Why does this happen? Why, with this incredibly rich amount of data, we cannot draw simple and credible conclusions in the last few centuries? It is very unlikely that scientists are stupid and don't understand their job. They are very intelligent people with extensive knowledge in this area. The simple and logical answer is that they are forced to do this!

The data simply has to be agreed, be unified and classified. Then have to make a chart with time scale and with a parallel analysis of all the simultaneous events and conditions – air composition, temperature, plants, bacterial life, sea temperature geological and chemical conditions. Analyze it and draw a logical conclusion based on all! (Not only on selected) Facts and information that we possess. Such a task is simple; it can be done by any group of 1st or 2nd year students. The fact that this is not done is a clear example of how fundamental scientific knowledge is suppressed and manipulated by religion and politics, because scientists are relying on public funding for their work, and you know well, who is in control of those funds

Scientists tell us that modern humans came into Europe from Africa around 35,000 years ago. The Neanderthals, - the native population of Europe for 200–300,000years, suddenly became extinct. The dark-skinned modern humans from Africa arrived in Europe and Scandinavia and in only 10–20,000years they miraculously obtained an extremely blond Scandinavian fair complexion! Is this possible?

Let's consider the available facts to see if the official theory is credible and correct. If we are coming from Africa, why we are not black skinned? How long a time is it necessary for a group of humans to change skin pigmentation color? Are we having real historical facts for the time necessity for skin pigmentation changes? The evidence suggests that we have real facts and available data to have a credible understanding of this subject. - The Native North Americans came from Asia. They crossed the Bering Strait about 20,000years ago when the Ice Age provided a land bridge between Asia and America. Their origin was from the Eskimo people in north Asia. The available data suggest that the Eskimo population of Asia had been there at least 25,000years prior to this event. Some evidence suggests that the Eskimo people had been there at least 45,000years. The Eskimo people and Native Americans to this present day are dark-skinned, with black hair and dark brown eyes. This is an undeniable historical fact, and it is obvious that in 45,000years the Eskimo people's skin complexion didn't change at all, despite that they are constantly living in extremely cold Arctic conditions. The aboriginal people of Australia populated the continent 50-60 thousand years ago, and despite this long period, their population of Tasmania, which living in a climate similar to European didn't change their skin pigmentation at all! I would like to consider those facts a bit further. We know that Scandinavians had been blond already about 10,000 years ago. If we deduct this number from proposed 35,000 years - this will shrink the period for possible changes to a maximum of 25,000 years (this is according to the official theory) - for dark-skinned humans coming from Africa to evolve into pale-skinned blue eyes blond Scandinavians. It is obvious from the undisputable evidence we possess that this is a very short time for such a transformation to have occurred. And this fact does not give any possibility for the official theory to be credible or be correct! Even the written record from ancient civilizations, Greeks and Roman, suggest that the majority of Europeans come from Asia, not from Africa!

What is other evidence there for the early human population on Earth?

As we discussed above, all the evidence suggests that the Eskimo people have been living in north Asia for at least 45,000 years. They are modern humans, and there is no evidence in their physical appearance to suggest that they had come recently from Africa. That means that modern humans have been presented in Asia long before the migration of Africa's modern human have occurred. The Neanderthals have populating Europe for a long time! Officially they are not recognized as modern humans, but the beautiful cave painting left behind revealing their advanced culture. The reconstruction of their physical appearance, putting them very close to us! Despite that the last ice age has covered the continent with a thick sheet of ice, there is evidence for a human presence on the European continent as far back as 950,000–800,000years ago in Norfolk, where the footprints of five individuals in sandstone have been found. Other findings are the Swanscombe Man (400,000 years ago), Boxgrove Man (500,000 years ago) and in Essex (400,000 years ago), clear evidence that humans have been on a regular basis on and off the British islands. The earliest Neanderthals present in the British islands is about 400,000, - far earlier that the official theory suggests.

Reconstruction of Neanderthal   Reconstruction of Neanderthal woman

The changing of the climate (Ice Ages) is a possible reason for their migration to Europe and back, but that means that in Europe, there has been a bigger pool of human populations. In Spain, in the cave at Atapuerca has been found 28 human skeletons dating to 400,000 years ago. The analysis put them closest to Denisovans whose origin was in Asia. Denisovans are modern humans; their DNA is present also in the local population of Greenland. The recent findings suggest that Denisovans were interbreeding with the Polynesian population of southern Asia and the Indonesian islands in the period as far back as 400,000 years ago. More evidence suggests that the Indonesian islands were populated 1.5 million years ago.

The physical and structural appearance of the Aboriginal population of Australia is very distinctive, different from a possible origin of recent African descent, but they are modern humans, and their presence in Australia has been dated back to 65,000 years ago. In January 2016 in Carpenters Gap in northern Australia has been found a stone axe dated 45,000–49,000 years old. This is the oldest found axe on Earth. This fact suggests that the Aborigines were well advanced, able to produce tools before anybody else and without the help of African "modern" humans. The second-oldest axe was found in Japan, dated 40,000 years old. That shows the ability of local people to be toolmakers, and be intelligent enough to be able to cross significant sea distances in big number to be able to colonize Australia, Japan, and all Pacific islands! And they had done this in time when the "modern" humans from Africa did not possess such advanced technologies.

All this evidence rules out the credibility of the official theory stating that modern humans came 35,000 years ago from Africa. Recent findings for early human presence in the Siwalik Hills in India is about 2.6 million years ago. Homo erectus is found in Damanisi, Georgia, dated 1.85 million years ago. Recent findings in China suggest that an early human presence up to 2.48 million years ago! But all those findings are findings of different primitive humanoid groups, which is very likely that have no direct link to modern human (homo sapience).

The palaeontologists in late 18 centuries have found numerous sites of early modern human activity where has been many broken and mark animal bones along with good quality stone tools. The Bolivian modern human footprints are dated to be 10 – 15 million years old! In Baltic amber is found sophisticated woven fabric, which is dated to be 25 -35 million years old! In Hungarian – Vetersszolos is found the presence of campfires, and there also is a modern human truck of footprints dated 500,000 years! In Savona – Italy have been found modern human skeleton dated back 3 -4 million years! In Delemont – Switzerland is found skeleton 30 million years old, at Midi de France another skeleton about 20 million years old! In America the modern human presence is found: -Sheguiandah Canada – 125,000 years; New Jersey - 114,000years; California artifacts 200,000 years; Mexico artifacts 250,000 years; Argentina scull 1 million years; Idaho human figurine – 2 million years! In California gold mine - skeleton and artefacts 30 – 50 million years. The dating of the different sites varies between 3million to 25 million years ago! To put under the carpet, such significant evidence is unacceptable scientific vandalism! Those findings and also some modern shape foot prints of distant past can have pointed in the direction that the modern humans have developed much, much earlier and co-exists with many groups of humanoid primates, which is not our ancestors! When the pseudo-science has established the incorrect theory for the human origin, the search for human remains and activity in the earliest periods has been seized. Any new finding which contradicts the established theory has been dismissed immediately, discarded, and not reported! – This is scientific vandalism!

All these findings are undeniable evidence that humans have been living around the entire globe for a very long period, the period measured by millions of years, not by thousands, how the official theory suggest! The people migrated and followed the climate changes, interbred and evolved in different places, but they definitely have not recent African origin, or were out of Africa millions of years ago! They have been smart, capable, social and peaceful people, able to survive climate changes and catastrophes, but most of them didn't survive the cruelty and brutality of the coming violent groups of the so-called 'modern humans from Africa.' The undeniable example of their cruelty is the disappearance of the Neanderthals at the time of their arrival from African.

Unfortunately, the evidence for this crime is written in our DNA! Yes, I am absolutely serious! It is written in our DNA! - We have up to 5% Neanderthal DNA, and from this DNA is missing the 'Y' chromosome, which is passed only from male. The existence of only 'X' chromosomes in our inherited Neanderthal DNA is unmistakable evidence of the brutality of the new arrivals from Africa!

It is easy to construct a picture of the events, where the male Neanderthal population has been killed, and the women have been enslaved!

Unfortunately, we also have inherited this brutality, which is the major obstacle for our progress, ethics, and survival as an intelligent society. (Recent example is the treatment of North American native population!)

Roman and Byzantium written records show that the majority of European nations – ancient Greeks, Gauls, Franks, Germans, Slavs, and other groups – came into Europe mainly in the period of 2–8 centuries from Asia. The majority of those people possessed fair complexion, which cannot be explained by an African origin.

Despite that the physical appearance and DNA of the European nations have been influenced by contact with the local populations, the mass invasions of Tatars, Mongols, Huns, Arabs and the Turkish European invasion they still possessing very distinctive physical appearance, which cannot be associated with African origin! Natural migrations have had a big impact on mixing DNA. The actual arrival of the European nations from Asia putting a big question mark on the theory that modern humans came 35,000 years ago from Africa. All those nations that came from Asia have very different languages, different cultures, and very different physical appearances from the people of African origin. Such diversity, and such obvious physical differences, cannot be explained with a simple arrival from Africa. Appearance of such different languages needs hundreds of thousands of years, not decades!

I believe that a possible reason for those theories is the well-subsidized effort of the Jewish super-rich elite to convince the world that the roots of human intelligence come from their land (where the migrating African population has settled temporarily for about 150,000 y). Such an assumption is not scientific,

It is baseless, and is not supported by the facts. Such theories have no evidential and scientific data in support and cannot be correct.

If we want to have a reliable picture for these events, it would be better to stick to archaeological facts, which cannot be denied or manipulated to a big extent. The locations and dating of these archaeological facts speak for themselves, and they don't need any further interpretation. The DNA information is a very good tool, but contaminations of samples and interpretation of the data – where hidden agenda can be easily inserted – makes this wonderful technique a very unreliable tool to draw a conclusion for the migration and origin of humanity.

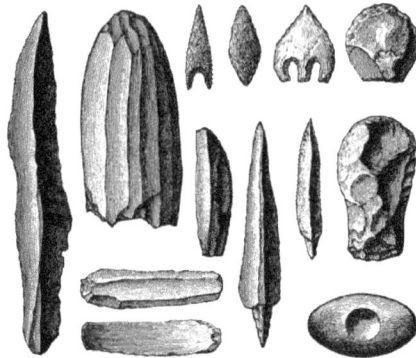

UNIVERSAL INFORMATION, CONSCIOUSNESS, LAW OF PHYSICS AND THE LIMITS OF KNOWLEDGE

We are living in a material world and our everyday experience making it hard for us to accept and recognize that in our Universe could be another form of fundamental structures, which are not material. In our material world, everything around us is made of matter and matter is just a concentrated form of energy. The universe is a very complex system, and to be able to understand it, we have to grow up and understand that the components of the universe not necessarily have to be made only from the known to us energy. They can be made from something else, physically different, which will make those components not to appear material, but they will be real physical elements, just will be in different physical form, which is not material. Those special non-material components of Universe are the universal consciousness and the quantum information. Our pattern of thinking, making it difficult for us to imagine how non-material elements can be real and fundamental in the structure of the Universe. To be able to understand and accept easier this concept, I will give an example of the other two well-known non-material components of Universe – Time and space. We have no trouble to accept their real existence as fundamental components of Universe despite they are not material, and we cannot fill or touch them. Similar as time and space are the situations with the consciousness and quantum information - they are not material, but they are real fundamental components of the Universe and are physically incorporated with the material structure of the Universe. They are the actual law and order of the Universe. They are ruling and regulating the physical processes of our world from the smallest particle, to the enormous universal structures.

Universal information:
What do scientists call quantum information? And why it is important?
In the late 19centuries and beginning of 20 century, the scientists have realized that all particles of matter and all universal structures are connected with informational link, which determines their properties and their interactions.
Einstein wasn't happy with this, because the informational link between the elementary particles is disobeying his assumption for the ultimate speed of light and travelling in an instant, no matter how far apart they are.
This link is called – "quantum entanglement," and Einstein calls it – "spooky action on distance."
The problem for Einstein does not stop there, because of the informational link between the biggest universal structures also traveling with instant speed! The evidence for this is the gravitational link between stars and planets. For example -the light from the sun is travel about 8 minutes to reach earth. If the gravitational link between the sun and earth travel with the same speed, the gravity will pull the earth in direction, where the sun was 8 minutes ago! For the distant planets, this effect will be much greater! This anomaly will create unstable orbits, and the planets will be thrown out, or will fall into the sun! Fortunately for us, the gravitational link (the information) between sun, earth, and planets is travel in an instant, and they know precisely where each other position is at any moment! Same thing we are observing in all universal structures.
The information for their position and physical state is travel instantly. The universal information is acting as a tool (or invisible hand) of the law of physics to implement its rules! The information is a fundamental property in the structure and order of the universe. The universe is functioning very much like a computer, where the law of physics is the software, and the information is the electrical impulses and link between all the parts of the components - (carry the information between all the components of the system)
The universal information is the order-keeping of the physical system; - it determines the size of each particle, the strength of forces, and the bond of atoms in matter. In short, the information is a crucial fundamental property for our world to exist and function!
Physics state that the information cannot be created cannot be duplicated, or

destroyed! It is important to know that the quantum information is protected with unbreakable code! - The information is created on the way to be 'stupid proof' with good logical reason!– (No-one is able to mess with the order of the universe!)

We have already started to manipulate and intervene in nature.- DNA allows us to do this and even allows us to destroy ourselves in this way, but the knowledge of how to intervene in the structure and order of the universe is carefully guarded secret!

The Information is embedded in the world in the form of quantum information in the matter of the physical world, and like DNA in the living organisms.

The importance, existence, and functions of universal information are carefully kept a secret from public knowledge, and the leading scientists avoid even mentioning or commenting on this subject, especially on its origin!

The quantum information poses and functioning with unimaginable sophistication beyond our best imaginations! From physics we know, that the pairs of particles are linked together with their own unique information link - (quantum entanglement). The problem is there, where the number of the particles in Visible universe is $10^{80,}$ and this is only 1% of the total matter, the rest 99% is diffused matter as gases, plasma, dust. When we take into account this matter too, the number of particles (only in the visible universe) is reaching infinity! But the problem is not stopping there, because we know that the universe is much bigger that the observable part, and this increasing the number of the universal particles to infinity multiply by infinity! In such unimaginable number, how possible is to be created and maintain unique specific band for the informational link between each pair of entangled particles? Our technology suggests that for each recipient must be allocated part of the wavelength of the radio transmission. This is a real and actual restriction for any communication transmission, and we the humanity are already struggling to find a way to provide more telephone numbers to satisfy the demand!

The sophistication of the quantum information is beyond the ability of the "Standard model" to explain it, but when using logic, we can get very good understanding of this phenomena.

Here are some facts that have fascinated the scientists a hundred years ago:

In the early 20th century scientists found some mysterious properties and behaviors of elementary particles. They found that fundamental particles which form matter are ruled by strange and mysterious laws - laws that are not valid for the macro world. They found that the particles of matter are connected with hidden information links, which determining their physical property and behavior. At the beginning of the 20th century, particle duality was accepted. That means that particles exist in two different states; as waves and as particles. The particles have different angular momentum or spin. Scientists found that when the particles are in a wave form, they never forget their specific spin, and when they return back in the form of a particle, they always revert to their original spin again. This is the fact behind the statement that quantum information cannot be lost. - The fact is that this information is stored somehow and somewhere, leading to the conclusion of the existence of hidden order behind the scene, which we called 'The Law of Physics". The current science believes that the information is stored in the matter! This is a very naïve assumption and shows their absolute misunderstanding of property functions. Such a scenario is impossible! - To believe that in every particle of matter is imbedded information storage. When the particle disappears where this storage will be? It is important to realize that the universe is constructed as one giant computer where the information storage is safely stored in universal consciousness far beyond our reach!

The other phenomenon which has been discovered is that particles are connected with information links - when you affect one of them, the connected pair reacts in an instant to restore the balance, no matter how far apart they are, even if it is light years distance! This interaction is called "quantum entanglement," and currently it is used as the basic function of quantum computers.

The third phenomenon related to our subject is the phenomenon that the wave state of particles exists in a state of probability, (uncertainty principle) and the particles two properties- velocity and position, are in strict relationship to our knowledge of them. - The increasing knowledge of one of those properties proportionately reduces our ability to know the other property!

This phenomenon acts as a very effective limit for our knowledge and ability to manipulate the property of matter. (with very good reason)

The fourth phenomenon is the so-called observation paradox, where the consciousness of the observer forces the wave state of the observed particles to collapse into real particles.

The double-slit experiment has been performed with one single particle at the time allowed to pass through the slits. When the path of the particle is not observed, those single particles passing through the two slits in the form of waves, when the path is observed, they pass through one slit only in the form of a particle. Even when the observation detector is on, but the data is not collected, the particle passes the two slits in the form of waves. This proves that only our conscious knowledge for the path of the particle makes the difference and collapses the wave function of the particles. This is a mysterious limit for our conscious knowledge of particles property imbedded in the physical law and property of matter. (Also with good reason)

Recently the double-slit experiment has been done in a sealed concrete and steel bunker, fully insulated from influences from the surrounding environment. Only through the internet can a single person on the other side of the earth see this experiment. When the person looks at the monitor, the particles pass through one slit only like particles; when he is not watching, the particles pass the two slits in the form of waves. Even when this person is not watching the monitor, but concentrates his mind and imagines the actual experiment, the results are the same! - His mind is influencing the experiment and the behavior of particles on the other side of Earth! This experiment has been done through government funding on a big scale, involving thousands of people, and repeated hundreds of thousands of times. The results are indisputable! The entire phenomena are beyond the classical physical laws for ordinary matter, but they are not a mystery (how the establishment present them), because they are functions of the non-material aspect of the universe – its software, or informational blueprint for the physical organization of matter. Exactly this is misunderstood from current physics! - <u>The fact that not the consciousness, but the information is collapsing the wave functions of matter.</u> This becomes obvious from the fact that the detector is an unconscious machine, but by gathering and storing information, it is affecting the property of matter.

The idea, or the 'blueprint' and the order of microscopic world don't have to obey the classical laws for ordinary physical structures on a big scale.

To be able to explain this subject better, I would like to start with our current laws for the protection of 'intellectual property.' Our society has a good understanding that intellectual property, which I will call for short (the Idea), is something real and valuable, and we are taking care to protect it.

What the Idea actually is? It is non-material phenomena in the form of a story, a melody, a book, a design, instructions, calculations or plans for something to be constructed or created. The Idea also exists in the form of computer program software. Our understanding for the Idea is that it is not material but is a real package of information, and is a product of an intelligent mind. <u>It is an instruction, or blueprint, for some sort of order or knowledge for something how to be created.</u>

The Information is the actual blueprint, the physical law (or software and program) for everything that we observe around us. It is in two different aspects and form – biological and physical.

DNA is the information and driving force for the existence of every living organism on Earth.

Quantum information is the major factor that organizes the transformation of the simplest physical substance - energy, into this amazing in variety world. Physics states that information cannot be created cannot be destroyed, and cannot be copied or duplicated. It travels with instant speed at any distance, and even back in time. In addition to this, I will add that the information is coded with an unbreakable code – something that we never imagined to be possible to exist – (unbreakable code), yet now the military is about to use it to make their communications absolutely secret. The coding of the information is a manifestation of intelligent logic embedded in the structure of matter that prevents us from destructive knowledge and the ability to interfere with the harmony of the universe. Raw nature is not logical and never creates conditions or embedded secret codes. – For example, DNA is an open book to be explored!

We are facing incredible logic, balance, special properties, and intelligence in the structure of the universal information.

The founders of quantum mechanics weren't naïve. They realized that all that amazing complexity of logical functions and incredible properties of information led to a logical conclusion for an intelligent origin of the information and universe. But before they were able to reveal to the world this great discovery, the news was immediately suppressed, twisted, falsified, and presented in a completely different way. Instead of the amazing truth, they were forced to present the public with a cover-up absurd theory where the act of observation creates the reality, and we are the creators living in an imaginary holographic universe where nothing is real!

To expose the nonsense of this, I can say: OK, if the observer creates the reality, then where does the observer come from and where he is situated?- To have an observer, he must exist in reality!

The voices of the few remaining honest scientists have been muted; we still can hear the echo of the Einstein statement:

'I don't have to look up to know that the moon is there.'

Even such an obvious example has been ignored.

To consider the universe without its informational organization is the same as to consider computer (physical appearance only), without taking into account its software and information processing functions! And still, the pseudo-science hiding and refuse to talk and consider the informational link as an active part of the dynamic physical organization of the universe!

The ruling tyrants and crooks are bombarding us with lies, conflicts, and confusion. They are taking away our knowledge and playing smart games with the most valuable things we possess – the knowledge of our existence, the knowledge where the sign of the creator is and what his actual manifestation and presence is! What his legacy, moral, and ethical conditions are, and what plans for our future are!

They are hiding crucial knowledge that in the DNA of a single cell can be written the knowledge of entire civilization. That on the unused parts - (junk DNA) the written messages can survive for billions of years unchanged, and or our, DNA is a clear sign of encoded language sequence, which could be the message to us from our possible creators!

Let's come back to the basics, - to the structure of matter and universe and the fundamental building blocks of matter. The crooked and fake scientists hide and refuse at any cost to reveal the real nature of the fundamental structure of the world. We know that those properties exist, and they are real and fundamental: matter, space, time, information, and consciousness. These are the known building blocks of our world, where the only material component is matter. The other four – space, time, information and consciousness – are not material, but they are more fundamental than matter because they are the cause, the organization, the law and order and the reason for matter to exist, to function predictably and to be able to form everything around us. Space and time, we will consider in a later chapter. I will not allocate much space to explain them.

The other two components - consciousness and information, are the main subject of this chapter.

<u>Consciousness</u>:

Consciousness is not a well-understood phenomenon. I believe that consciousness is bond, or "symbioses" of information, intelligence, and some form of unknown for us (nonphysical) energy. It acts as the "hardware" of the universe, which stores the information and the law of physics. It also is a symbol of life, because, without consciousness, even the most powerful computer is just a lifeless machine. From quantum mechanics, we know that consciousness is more basic and fundamental than information because consciousness affects and influences information and the outcome of some scientific experiments, but we never observed information being able to affec or control consciousness. The fact that we, humans possess consciousness gives us some understanding of its real origin, functions, and properties. - Our mind is using a small part of universal consciousness. The mind and consciousness are incorporated, and that's how we have a conscious mind. Our mind is controlling all the processes and functions of our body. It is not difficult to realize that the universal consciousness has similar functions -producing and storing the informational order of the Universe and controlling the physical processes in the Universe. We are witnessing the order, where consciousness is the phenomena, which creates the law of physics, universal information, and the Universe by self!

<u>It's easy to understand that information can be a product only of a conscious, intelligent mind! Pile of dirt or lifeless matter cannot produce information! Same is valid for an explosion – Big Bang cannot produce sophisticated logical order!</u>

This realization is actual undeniable evidence for the presence of the creator of the universe, because according to quantum mechanics, for the universe to exist, it needs a conscious observer on a universal scale. I don't have to specify "who" this intelligent observer on a universal scale would be! I just would like to explain that we are facing the actual mechanism of how this world has been created!

Information cannot spring out spontaneously from nowhere; it is an intellectual product of universal consciousness. It is embedded in the fibers of matter and is the tool and mechanism of the law of physics, to keep in perfect balance all physical processes. In the properties of information are embedded very logical principles for the structure and stability of matter, its functions and the purpose are everything in the universe to work in balance and harmony.

Max Planck – the father of quantum mechanics, states:
'I regard consciousness as fundamental. I regard matter as derivative from consciousness. We cannot get behind consciousness. Everything that we talk about, everything that we regard as existing, postulates consciousness.'

Max Plank

(If we are intelligent enough we will understand that he is telling us where the universe comes from! – (Exactly this knowledge has been hidden from us!)
The consciousness is mystery and puzzle, which always have fascinated human been. The ancient philosophers have tried to give us an explanation; the religion associates the consciousness as our spirit. The current scientific view is that this is a puzzle beyond our knowledge and ability to understand.
Is really the understanding of consciousness being beyond our knowledge, or as usual, there is the reason for the elite to hide the answer of this puzzle?
The disturbing facts are that in the early 20th century the scientists have found the answer of this question, but with the scientific cop of so-called "Copenhagen Interpretation" which they adapted in physics, the truth has been suppressed very quickly and effectively!
The voices of the great honest scientists have been ignored!
Max Planck – the Nobel Prise winner and father of quantum mechanics state:
"I regard consciousness as fundamental. I regard matter as a derivate from consciousness. We cannot get behind consciousness. Everything that we talk

about, everything that we regard as existing, postulating consciousness."
Is anybody can state, that the father of quantum mechanics doesn't know what he is talking about? How such a vital statement as: "I regard matter as a derivate from consciousness" get unnoticed? - It is reviling where the universe comes from! Are we so stupid not to understand the importance of this simpl revelation?

How such vital information as – "I regard consciousness as fundamental"! Is not considered and study by the current science?

Why is the current science giving us such bizarre explanation of consciousness where the consciousness creates a fictional world as hologram where nothing is real and exist only in our mind? Is any normal person can accept and believe in such nonsense? Are they explaining in this scenario where our conscious mind comes from and where it is situated to create their fantasy? How the sciences get such low to this primitivism?

As usual, the elite sponsored pseudo-science not giving us any credible answe and creating mess and confusion, but do we have to accept it?

Let start consider the available scientific facts, analyze them with care and logic in order to find the truth behind this fascinating subject – consciousness. The fundamental mistake in consideration of consciousness is that most of us believe that consciousness is a product of our brain. And as usual, in order to create misinformation, this wrong concept is wildly promoted by the pseudo-science! But this is absolutely incorrect, and I will try to explain why:

- The consciousness is present in every part of our world – it exists is in every part of the micro world and atomic particles of matter.
- Consciousness exists on a grand scale and is part of physical law and order of the universe. You cannot create, destroy, or duplicate consciousness! (This is valid for quantum information too)
- The recent scientific findings of the property of water have reviled that the water has a memory and carry information about everything it touches. The shocking discovery is that the water reacts and carries information of the consciousness of the person who gets in contact with it!
- There is overwhelming evidence that consciousness exists in every living organism – from single sell to the biggest three of the jungle.
- Well-known also is the existence of consciousness in the animal worlc

The most crucial fact for our understanding of consciousness is the fact that the plants have consciousness without having a brain! – The consciousness exists; it is everywhere and is not a product of our brain!

When we are starting our life as a union of two sells to form an embryo, our first two sells already possess consciousness (or self-awareness) other ways; those two sells will not find each other! Long before we develop our nervous system and brain, the consciousness is already there! - We are given consciousness!

We have to understand the fundamental principle of how our conscious mind is build and works! Our brain is like a computer –is our tool for processing information. Like every computer, our brain needs the basic software to start functioning, and this software is the consciousness! To become self-aware and to be alive. First, you need consciousness! What happens with the consciousness when our brain is dead, simply I don't know. The fact that the consciousness exists independently of our brain makes me believe, that the end of our brain is not the end of our consciousness, but how it will continue its existence - as an independent package or will merge with the universal consciousness is a mystery and is question beyond our understanding.

I would like to continue the consideration of the facts with the aim to shade more light of this most important phenomenon of our life:

Why do Max Planks believe that consciousness is fundamental?

In the previous part, we have considered what the hierarchy of the known property of the universe is. We get to the conclusion, that the (quantum) or the universal information is the mechanism, which the law of physics using to implement its rules. We have reached the conclusion, that the law of physics is actually a package of information as a program, or set of rules for the matter and universe to be created and to work in harmony. It is more than obvious that these packages of rules and information must precede the existence of the physical universe in order for the universe to forms and exist! It is obvious that those programs and information have to precede the universe and must be created by an intelligence, or intelligent mind! The row matter cannot exist without informational organization and is obvious that the information must precede the existence of matter! And there in focus coming to the consciousness! It is the answer to the puzzle! - The consciousness is the phenomenon, which is responsible for the creation of the law of physics and the informational organization, which is a link between all elementary particles of matter in the universe! In short, consciousness is the first fundamental block of everything around us to exist! Is obvious and logical that consciousness has created the universe!

It has created us also! And the result of it is amazing! We are observing the harmony of the universe, where in every part of matter and nature is imbedded consciousness logic and intelligence! - This is the knowledge which the elite don't want us to know and is suppressing and muting the science, which relies on government findings!

I am sticking strictly to the scientific aspect of this phenomenon, and I am not willing to go in mysticism or use a mystical or religious explanation for the origin of the universal consciousness. I am using only logical consideration of the available scientific facts in aim to find the correct answer to this vital question.

We are very lucky to be able for the first time since the humanity exists to be able to have a credible scientific answer of the puzzle how the universe has been created, where the law of physics come from, what the consciousness is and where the biological life comes from. Still, the origin of consciousness remains elusive and unknown, and I have no other choice than to stop short of comments and explanations because I am not willing to start baseless speculations about the origin of the universal consciousness. It is beyond our reach, it will remain puzzle, and probably we will never be able to find the answer where it comes from!

I just would like to come back to consideration of consciousness, because we can find its presence in every part of nature and in every living organism. The consciousness is really something very special, which giving a unique quality to every living organism!

To have a better understanding of the world, we have to start with the basic understanding of this phenomenon:

What actually is consciousness of living organism?

On the base of the wrong assumption that intelligent thinking is a sign of consciousness is based our incorrect understanding of the world! This incorrect assumption is proven by the current effort of some environmental protecting groups to spare the life of highly intelligent mammals like whales, primates, and dolphins. For their surprise, I would like to explain that consciousness and intelligent thinking are two very different things! Actually, consciousness is the real cause for the appearance of intelligent thinking creatures.

The consciousness of the living organism is self-awareness in the surrounding environment! Many people are mixing those two terms – Intelligence and consciousness. The capacity of the brain is determining how much consciousness can possess and processed.

I would like to explain the difference! - Intelligent way of thinking as a problem solving, or logical information processing the supercomputer can have, but no matter how sophisticated is this supercomputer, the computer will always remain lifeless machine and will never possess consciousness!

The consciousness is giving us our special quality of fillings and emotions! Our fillings of love, dedication, curiosity, humor, music, loyalty, self-awareness, and ethics! - This is the special gift of consciousness, which make us different of the lifeless matter, or intelligent machine! We have to understand that we are connected to the consciousness of nature and to universal consciousness! We are part of it! We are part of nature and the universe, where everything is created with logic and purpose, and we also have to know how to find our own purpose!

Many people also believe that consciousness is quality only of the advanced (Intelligent) organisms of the natural world and that the more primitive forms of life not possessing consciousness. They believe that the presence of brain and neurons are the key necessary elements for consciousness, but they are wrong! Absolutely wrong!- Completely wrong!

We have to develop an understanding that all living organisms are possessing a consciousness, and if we have to be honest, we have to admit that all living creatures have the same right of existence! It is beyond the ethical standard of intelligent creatures (us) to destroy the nature and slotter millions of animals on a daily basis in order to eat them! Is time to use our knowledge and technology to start producing our food without the necessity of killing the others!– The basic elements of our food are carbohydrates! This is exactly the structure of natural oil, gas, and even coal! We just have to transform them in an appropriate for us form!

Let see how far the consciousness is imbedded in the living organisms and is the simple bacteria and plants have consciousness.

I will start with the so-called primitive bacteria's. - To be able to survive, all bacteria are aware of the surrounding environment; they are able to navigate, to sense and find food or avoid danger, to react on light, chemicals, communicate with each other and using defensive or offensive chemical warfare. They can learn to avoid some hazards – (have memory). All those quality and functions of the simple bacteria's are exceeding greatly the basic requirement of the formulation that the consciousness is self-awareness in the surrounding environment!

The most interesting results are that the plants have complex emotional and censoring mechanism based on consciousness. By using sophisticated censoring equipment, scientists have found that the plants react to our tot's, emotions, stress, and are emotionally bonded to the person who is taking care of them!

The study of Cleve Backster has recorded that the plants react instantly at the moment when the life of their carer is in danger. The most important experiment he did, is when he placed plants and recording equipment in a crater in the desert and pointed them to the sky in aim to use the plant consciousness to find the existence of biological life in the universe. In regular intervals in phase with the movement of some celestial bodies the plants have indicating a connection to the consciousness of biological life of the universe. The results of this experiment are controversial and are not recognized by the scientific community, but the great achievement of Backster is this, that he shows us the way how to use consciousness to start looking for conscious life in the universe! We can understand the true extend of consciousness that is in every part of matter, every atom, bacteria, plant, and every living organism! That's why the father of quantum mechanics Max Planck has stated:

"I regard consciousness as fundamental"

Because it is the real foundation of everything!

We have to take seriously the statements of such colossal figures as the founders of our current knowledge because those peoples have proved their deep understanding of the nature and structure of our world!

We have to understand that in the structure of matter is imbedded mechanism for the physical interaction of our consciousness with the surrounding world, which gives us freedom of choice, but the uncertainty principle makes the outcome of our actions uncertain and unpredictable. On this way, we are given freedom of choice. We are also given the principle of ethics, which are a privilege only of intelligent minded creatures! The sad thing is that in the law of physics is also embedded an ethical barrier or critical ethical limit for the survival of civilization! Currently, we are on the edge of this limit, where our technical abilities are advancing faster than our mental capability and ethical standard. It is sad because technical advances in combination with mental, moral, and ethical degradation is a real recipe for disaster! – It is a similar scenario as to give a loaded gun to a small child to play with.

It cannot be an accident that we are given a beautiful planet, freedom of choice, endless resources, intelligence, and at the same time there is imposed a strict limit for determining the future, and a limit for destructive knowledge on a universal scale, and a limit for instant communication and fast space traveling. All these conditions and limits imbedded in the physics and property of the world show an absolute logic and understanding of our potential cruel nature! - Understanding what our corrupt destructive and egotistical society is capable of doing to the other civilizations.

Is sad to write this, but this is the correct description of our social principles, where the tyranny is labeled as democracy, the exploitation as cooperation, the invasion as liberation, the lies are mass information, and orchestrated deception is labeled as science, and where even religion is used for political purpose and for control of the population!

We are given choices to correct ourselves and become ethical, but if we choose to continue live on same corrupt and destructive way where our further technological advances will become a threat for other civilizations, the appropriate measures to prevent this scenario are already in place!

A conscious observer on a universal scale?

The logic behind this order is something which raw nature cannot create. Such a logical order can be the product only of super-intelligence with a good understanding of how corrupt an intelligent society could be! It is obvious that the set of preventive conditions are not negotiable! - The quantum Information one day could be our greatest tool for instant communication with other civilizations, but first, we have to survive the ultimate test and be able to go through the ethical barrier for advanced intelligent society.

We have to start thinking: do we have to continue our arrogance to ignore and oppose the universal principles, or do we have to start adapting to them and putting an order, ethics, harmony, and sense into our life?

Let us consider what we know about this mystery. The official version for 'Dark Energy' in short is this:

Observations show that all the galaxies of the universe are retreating away from us at a continuously increasing rate of expansion proportional to their distance to us. In the first 7 billion years, the universal rate of expansion was slowing down in accordance with the laws of physics, logic, and our knowledge. But something very strange happened when the universe reached 7 billion years old. The rate of expansion started accelerating with continuously increasing speed so that the furthest galaxies ran away with greater speeds proportional to the distance from us. This is a puzzle, opposite of any logic, the laws of physics and our best understanding because the gravitation pull of the galaxies and matter in the universe should continue to slow down the expansion and we know that the earth is not the center of the Universe!

Scientists have calculated the amount of matter responsible for such huge negative gravity. It comes as a big surprise and shock to learn that another 72% of matter is invisible, with an unknown origin, unknown properties, and an unknown location. There are various proposed theories, but none of them can give a credible explanation of this mystery. Science is not able to detect this mysterious matter, nor to have any credible explanation for the origin of this increasing strength of negative gravity. The well-established opinion says that that our current laws of physics cannot explain and solve this mystery. The established opinion – that the missing matter responsible for this phenomenon is evenly distributed through space is fundamentally incorrect because in this case, the negative gravitational pressure (force) would be equal in all direction, equalize itself and would not move the galaxies at all. But in our case, the force of Dark Energy is acting in one direction only! - This can be explained only if the responsible matter is unevenly distributed and is denser closest to us. This is very unlikely because the Earth is insignificant and is not the center of the Universe. Also, in this scenario, the model is fundamentally unstable, because when you disturb the perfect smoothness, matter starts clamping and collapsing. In our case, with "negative" gravity will be the opposite: this invisible matter will have a tendency to spread, equalize the density, and become smooth. We are dealing with matter 14 times denser that ordinary matter and this matter has time, from 6 to 13.7 billion years, to spread and equalize its density and become smooth. Obviously, this didn't happen, and definitely something is not correct with the official assumption, because the ordinary matter is supposed to be only 4% and manages to clamps together and form stars in about 400 million years after recombination. This is a credible example that such a scenario for this dense, mysterious matter cannot be correct. All this fact is leading to the conclusion that we don't understand something or our observational data is incorrect.

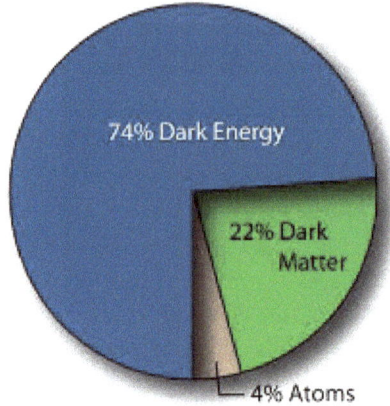

The proposed structure of the universe (courtesy of NASA)

Let us consider the three main attempts for explaining this mystery:
1). Vacuum fluctuations produce negative gravity and, with an unknown rate of distribution, push the galaxies away. This explanation is just hypothetical, with no basis of observation, proof or accordance with any laws of physics and has a wrong unworkable model of forces distribution. We have no other option but to ignore this explanation.
2). Space is expanding, But this expansion in order to match the observations must be an uneven spatial expansion, which is not the case because space in our Universe is uniform and our Universe is flat. We know this with certainty. If there were an uneven rate of expansion of space, we would observe a significant bending of light when it traveled between distant objects, which definitely are not the case! The other problem with this explanation is that there is no experimental evidence and no any proof that space really can expand. This also will violate the laws for conservation of energy. This facts is tell us that we have to ignore this explanation also.

3). There is invisible matter interacting only with negative gravity; How I stated above, to accept this scenario, we have to accept an unworkable model of force distributions, and matter with absurd properties contradicting the laws of physics and all logic of accumulated knowledge. From the Theory of Relativity, we know that space, time, and gravity are related and form the fabric of the Universe, or the total amount of energy in the Universe is constant. That means that we can put them into a simple equation; +Space x +Time = +Gravity but to have a negative value for gravity, we have to change the value of one of the other components of the left side of the equation. We cannot put negative values on space, so the only option is on Time. If we put a negative value on gravity in our equation, the negative value of gravity actually will reverse the direction of Time! The equation will look like +Space x −Time = −Gravity. That means, in order to have negative gravity, we have to change the direction of Time, which we know that is not the case! Our Universe is a closed physical system. We know that we cannot put same properties with opposing values in a closed physical system, because they will cancel each other, like matter and antimatter, cold and hot. The same is valid for positive and negative gravity and positive and negative Time. We are considering the existence of matter 14 times denser that ordinary matter, and if such matter is there, it will cancel our gravity, will reverse our Time, and will destroy our Universe! It is absolutely obvious that negative and positive gravity cannot exist in the same Space and same Time! - This is absurd, and this absurd explanation has to be ruled out!

After examining the facts and considering the laws of physics, the conclusion is that all three explanations are baseless and cannot explain the phenomena of the scenario of "runaway universe". So... where is the mistake?

The Standard Model is based on the assumption that the Universe exists in a void and is spherical. In this case, there will be no problem if we accept that the Universe is spinning also. Everything solid is spinning in one direction because inertia produces a gyroscopic effect. But the Space of the Universe doesn't have to obey this rule, because it is not solid, and the gyro effect doesn't apply to it. If we understand this and apply this understanding that free-floating Space of Universe can spin in one or two directions simultaneously then everything changes. The jigsaw puzzle falls into place with unimaginable precision and beauty. - The answer is simple: Space spins in two directions, where the axes are 90 degrees to each other. Centrifugal forces pushing the galaxies away from the center, and the increased radius provides the ever-increasing runaway speed. Currently, this is the only possible model that fits the data presented and explains this mystery completely without violating the laws of physics. There cannot be any mysterious energy or missing matter. The calculated 72% missing matter cannot exist because positive and negative gravity cannot exist simultaneously in the same space and time. They will cancel each other!

The failure of all scientific experiments to find any local negative gravity, any matter or particles responsible for such an effect is current proof for the correctness of my assumption that such negative gravity simply cannot exist.

I have proposed this solution to the Astronomical Society, but for some "mysterious" reason, and without stating or finding anything wrong in my proposal, the Astronomical Society rejected it and preferred to stick to the unscientific and absurd explanation of this alleged phenomenon. I was puzzled by the absolutely illogical and unscientific behavior of the Astronomical Society and went further to study the phenomenon. What was my surprise when I learned that I am not the first person to have discovered the nonsense of this newly discovered phenomenon or proclaimed "mystery". Even some respected astronomers have paid a heavy price for expressing doubt and speaking out against the established official version.

It is a simple and undeniable fact that the assumption for the recession of the galaxies is a wrong and false assumption! The tool for this deception is the use of the so-called Hubble constant, which takes into account the light red shifts us a sign of the movement of the galaxies, but in this assumption, is not taken into account the other two mechanisms, which producing light red shift - which are the gravitation field of the Universe who depleting the energy of the travelling light and the electromagnetic field of the Universe, which producing the same effect. As a result, the light is losing energy due to interaction with the gravitational and electromagnetic fields, the light starts to appear as red shifted, but this in no circumstance could be interpreted as the recession of the galaxies! This is simple, obvious, and is proven even by the Theory of Relativity and countless observations and experiments. The hiding of such vital facts, clues, data, and evidence is a shame for the people who call themselves scientists. The reality is that this phenomenon has been invented to cover up the nonsense of the Big Bang Theory and its inability to explain the pile of new observational evidence which disproving this theory completely! That's why the Astronomical Society doesn't want even to bother discussing reasonable questions or elegant solutions!

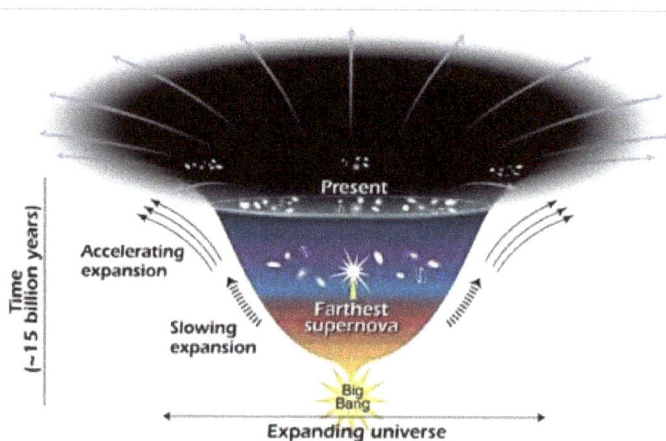

Dark Energy or dark fantasy? (courtesy of NASA)

THE MYTH OF DARK MATTER; KEPLER PLANETARY MODEL
AND GRAVITATION BOND IN THE GALAXIES

The puppet scientists of the establishment present us with another fantastic story for the existence of mysterious invisible 'dark matter.' The necessity for the creation of this mystery comes from the inability of the pseudo-scientists to hide the lie and impossibility of the Big Bang Theory. What happens?

Everything was OK until telescopes became more powerful and were able to see much further than 'they' liked. The astronomers were able to see further than 12 billion light-years away from where, according to the Big Bang theory, must be just primordial state of the universe, - just gas and dust. The unpleasant surprise was that where there must be an undeveloped universe full with only primordial stuff, the astronomers found a fully developed and mature universe with second-generation stars, contain heavy elements. Even with high exposure, a galaxy was found 15 billion years away. This was a big shock for them, and they quickly rushed to cover it up. The puppets scientists invented some computer simulation, and they denounced quickly the discovery of this galaxy 15 billion years away by 'shrinking' the real distance to the galaxy on their computers to what was comfortable for them – 12.5 billion light years distance.

OK, they managed to do this, but the other problem still remains! - How then, where there must be just a developing universe, is a fully mature universe with second-generation stars? (The first stars must have only hydrogen and helium, but the second-generation stars have heavy elements). To cover up this, they decided to invent something to disguise the obvious nonsense of the Big Bang theory. Yet, though these guys are very intelligent, they weren't able to produce anything convincing and credible. They started using mystics to cover up the nonsense. In order to explain why, where there is supposed to be only gas and dust, there is a fully developed and mature universe; they invented 'dark matter' – the mysterious invisible, undetectable, unpredictable substance, which helps the baby universe to develop 'on steroids' very quickly. The problem is that we cannot see this dark matter and cannot detect it at all! It contradicts all observational data, physical laws, and common sense, but the smartest guys insist that this mysterious force is real and must be there. Do we blindly have to believe them? Is this religion or science? How have they justified this absurd and mystical claim? - Very easy, as usual with lies and deception, and not allowing real scientists to have any publications or public comments on this subject.

How do they justify the existence of this fantasy?

I will start with Kepler's law for planetary movement. Kepler calculated and formulated his equation for the dependency of the distance to the sun of the planets in the solar system and their rotation speed. The equation is simple and relatively accurate. The catch-22 is that, in the solar system, 99.9% of the mass of the system is located in the center. The sun has 99.9% of all mass in the solar system, and the mass of the planets is only 0.1%. In this scenario, the mass of the rotating planets is insignificant and irrelevant.

Kepler's equation is relevant to the solar system. What about the rotation of the galaxies? Can we apply to them the same formula that we apply to the solar system? - Definitely not, because the mass of the galaxies is relative evenly distributed in the disc of the galaxy. There is assumed to be a black hole in the center of the galaxy, but it is not more than a fraction of the total mass of the galaxy. The evenly distributed stars in the disk of the galaxy are close to each other and form a gravitational and electromagnetic bond. This attractive bond acts like glue and forms one interconnected structure. The galaxy starts, and plasma filament is behaving like one rigid body. That is what we observe! Nearly all parts of the galaxy are rotating together because they are connected with an internal electromagnetic and gravitation bond. This is logical and absolutely obvious. We cannot apply Kepler's law, designed for the solar system where all the mass is in the center, to a structure where the mass is nearly evenly distributed, and the parts have an electromagnetic and gravitation bond. It is simple, logical, and obvious!

Compare the dense body of a galaxy and the empty structure of the solar system

The rotation of galaxies has been accepted, and nobody has seen anything strange there until the smart guys intervened and, with the aim to justify the nonsense of the Big Bang Theory, they declared that this is a 'big mystery.' What actually is their mystery? - They applied Kepler's law for the rotation speeds of the planets in the solar system to the galaxies and invented their 'mystery.' They start claiming that the galaxies are rotating very fast and the stars must be flying away. And because this has not happened, there must be some mysterious invisible and undetectable matter to hold them together. Fantastic, well done. Congratulations! All the scientists with common sense were shocked, but the well-lubricated machine for suppression silenced them. What happened, how come that madness has prevailed to such an extent? They have invented a mystery, and now the public has to provide them with further funds, big salaries and expensive facilities to study this nonsense? How did we get to such a situation, where the lies and deception have infiltrated every part of our lives? I want to analyze this situation because it is an extreme example of hidden corruption and arrogance! If you start asking these guys legitimate questions, they have a policy never to answer or comment on your questions.

If somehow, in a public lecture, you ask some questions, they philosophically will tell you that things are very complicated and are beyond your knowledge, intelligence, and your ability to understand! In real terms, they are telling you that you are stupid and you are allowed only to provide them with funds, big salaries and to keep quiet. Let's call the things by their real names:

If somebody deliberately deceives you and lies to you with the aim of extracting your money for his own benefit, this is called extortion. It is a criminal activity, and the person doing this is a criminal and must face justice, pay back the money, and go to jail. Is there any difference between this case and the situation with so-called legitimate scientists? Why are we tolerating this situation and not stripping them of their privileges and funds, because their activity is not of any benefit to the public or science? There are many other areas where those funds could be well spent.

This is the real story how "they" have invented the 'mysterious' dark matter, which is invisible, undetectable, not interacting with matter, has no known origin, has no known building particles, doesn't absorb heat nor radiates anything, does not clamp together to form solid objects, but provides only the "necessary gravity" for the smart guys to justify something more absurd and bizarre even than this substance – that this huge and enormous universe can be compressed in a dot smaller than an atom!

Well done! Such a manifestation of arrogance and ignorance must be awarded not with one, but with a bunch of Nobel Prizes and thousands of medals for dishonesty!

The tragedy of our time is that the nonsense and arrogance have prevailed and become our everyday reality. Some people accept this as normal and inevitable and don't even want to be bothered to hear or accept the truth. We cannot continue in this way! This is a dead end for humanity. We have to do something to put an end to this situation. It won't be easy because the biggest obstacle is our acceptance of nonsense and corruption. There is a way how this can be done, very elegantly and without revolution or violence which I will explain it in a later chapter, but we have to start with the rejection of nonsense and lies.

BIZARRE THEORIES IN ASTROPHYSICS

All the following theories are a manifestation of supreme arrogance, ignorance and have nothing to do with the real world. They have no any scientific value.

String theory

This theory is based on pure fantasy and mathematical manipulations in a bold attempt to produce a "Theory of Everything", but the tragic result is that they have produced a theory of "Nothing". The basic assumption for the development of this theory is that the Universe is constructed from incredibly small one-dimensional vibrating strings. This assumption is not based on any real observation or experimental clues. There is no such thing as a one-dimensional physical object in our three-dimensional world. Even the authors of the theory cannot describe how a one-dimensional physical object looks. It is absurd, just a pure assumption, without any real basis for it, but the one-dimensional strings are very comfortable for their further speculations and fantasies. These strings are also comfortably chosen to be the length of the Planck constant (10^{-35} – to fit the Standard Model). There is no such thing as a one-dimensional physical object! The length of the strings, which supposed to be physical objects, can be measured in two opposite directions from each end, which mean that they are at least two dimensional! - Only Time is presently known as a one-dimensional phenomenon! Time has only one direction, but Time is not a material object!

In reality, the String Theory is not only one theory, but many. In recent times, the authors of this hypothetical Sscience" have managed to reduce the main theories to just five. The pseudo-scientists have let loose their fantasy, and with complicated mathematical exercises, they have produced amazing things. One of the theories, the 'Bosonic,' predicts the existence of 26 dimensions. So far, currently science know the existence of only three dimensions.

Another theory - 'Super-gravity,' is more modest and restricts itself to only 11 dimensions. These theories even manage to calculate the "existence" of the other fantasies – Dark Matter and Dark Energy. Amazing "achievements" isn't it? I don't want to waste your valuable time with the useless details of those theories. I just would like to give you the final result of this shameless nonsense. The String Theory manages to calculate and predict nothing. - The result of their mathematical fantasy is incredible! They come up with a statement that the number of universes is 10^{500}, which exceeds greatly even the number of atoms in the visible universe which are 10^{80}! This number is close to infinity, which means that any result for any event or anything–any number, any force, any value or anything – must happen in one of these endless universes.

In short, they are predicting everything! - But non–specified "everything" is equal to nothing! And in this situation, they have the arrogance to claim that String Theory is the leading candidate for the Theory of Everything! - Without the ability to predict anything or to ever be tested, such claim can make only incredibly stupid or incredibly arrogant person!

I will quote the view of leading scientists about String Theory: - Peter Woit said: 'The String Theory is unhealthy and detrimental to the future of fundamental physics. The extreme popularity of String Theory among physicists is partly a consequence of the financial structure of academia and the fierce competition for scarce resources - (wasting of the public funds). The fact that even leading physicists like Roger Penrose are breaking the code of silence and expressing similar views proves that the nonsense and arrogance becomes unbearable even for the members of the "club".

The scientific standard for a theory to be accepted as a theory is that the theory must be based on real facts must be able to be tested and must have an accurate prediction of testable phenomena. So far, String Theory does not qualify as a scientific theory at all; it is just another baseless fantasy and scientific speculation.

Multi-verse Theory

This is officially recognized as a 'hypothetical speculation' which will never be able to be tested. Most of the leading physicists disagree and distance themselves from the idea and concept of this theory and reject it, or don't want to talk about, but the media love it! What actually is the idea of this theory? The reason for this speculation is a boldly arrogant attempt to please their masters and disprove the evidence of the intelligent design of the universe, which they provided themselves within the theory of the Big Bang. The mystical concept of the Big Bang and the majority of 'evidence' incorporated there strongly points toward the intelligent design of the universe. The first assumption that the universe has a beginning and comes out of nothing for no physical reason is absolutely mystical and can be explained only with the presence of God, but not by physics. After that, about 25 fundamental physical constants have been found to be tuned with precision up to one to trillions. By any standard of logic, this is more or less undeniable proof that our existence cannot be accidental, but is the product of careful, intelligent design. When "They" realized that they went too far in this direction, they started to worry. And as usual, the next step for them is to create confusion and doubt. And this confusion and doubt are exactly this – the creation of the theories of 'Multi-verse' and 'Consciousness creates the reality.' One of the sources for the Multi-verse theory is the String Theory; - we are witnessing how one nonsensical argument leads to another!

To deny the intelligent design origin of the Universe, they say that out there are billions of Universes, and it happens by chance or by accident that we are in a good one fine-tuned for life! There is no even a clue for it, and the impossibility of this scenario exceeds any imagination! Those two concepts are absolutely baseless, unscientific mystical and absurd, but despite the obvious incorrectness, they had huge publicity. The reason for which this mystical speculation has been created and has been promoted as "science" had achieved its purpose for public confusion.

In the previous chapters, I revealed evidence that the Big Bang Theory is impossible and false, but "They" prefer to stick to any nonsense that originates from them. It is absolutely typical for the establishment to produce such confusion. On one hand, the Big Bang Theory proves with absolute certainty the intelligent design of the Universe, but on the other hand, the crooked scientists bluntly deny all the evidence which they have produced by themselves.

Here is the next one:

Consciousness (or the observer) creates the reality, and we are living in a holographic Universe:

This statement, assumption, or theory came to the public's attention for the same reason – to create confusion. This is pure speculation and arrogant manipulation of the results of the famous "double-slit experiment", and has nothing to do with real physical events and everyday reality at all! The basis for this speculation is the effect of Consciousness on the wave function of the particles, but we also observe such an effect when the waves interact with physical objects.

No more, no less, this effect shows the universally embedded limit of our knowledge and the limits of our ability to know simultaneously the complete properties of particles - (their position and velocity). Our weak consciousness has very limited influence over the particles involved in our experiments, and there is no sign that we can create anything with the power of our mind. This theory is an extreme fantasy, denying the existence of the Universe and the real world and claiming that everything we see, experience, or feel is just a virtual creation of our mind! I will submit only one argument against this, to reveal the absurd nature of this nonsense: if the observer creates the reality, then where does the observer come from? Who creates the observer? And where the observer is situated? - Further comments are not necessary.

Wormholes

The fantasy of wormholes and short cuts through space and time is another form of extreme fantasy, based on scientific arrogance with no real evidence for support.

The assumption of the Theory of Relativity – that gravity does not exist but bends space – is obviously incorrect! We don't know what space is and the fact that we don't know its physical properties does not give us the right to make fundamental assumptions that we can bend it. Anyway, wormholes are an assumption that the extreme gravity of black holes will bend space and make wormholes between two parts of the universe, or time, or a passage between two universes. Then, for some mysterious reason, the black hole will disappear, but the wormhole will conveniently stay there to provide us with a free passage. This fantasy suggests that it will provide us with a way to travel faster than light, traveling to the past and to the future. The problem is also that space is three-dimensional, but wormholes are providing passage between two parallel two-dimensional sheets of space. - Well done! Brilliant! Even this is not an obstacle for the speculators. They just connect everything indiscriminatingly with their nonsense – space, time, two- and three-dimensional space, black holes and singularity and multi-verses! I don't have to disprove these fantastic scenarios, because everybody knows that they are pure fantasy and that such wormholes do not exist and there is no evidence in support of it. So far, even the existence of black holes in the form in which the officials predicted is not certain, because we haven't observed yet any black holes directly, and the current smaller distance between particles allowed by the law of physics can be only the Planck constant - 10^{-35}, (which is ruling out the assumption for the singularity of black holes). The speculators of black holes are ignoring the existence and validity of this constant and crashing the matter in singularity!

(As usual, they disobey the law of physics, but calling themselves physicists!) They are obeying the laws of physics only when the laws correspond to their assumptions, but when the laws of physics rule out the validity of their assumptions, they just ignore it, because they are very intelligent and are above the law! Unfortunately, we are paying them nice salaries to create this nonsense! The wormholes are good for movies or science fiction scripts, but they shouldn't be presented as a scientific theory.

Sun, Stars, and Supernova

Sun:

Mainstream science is giving us a Solar Model with an impossible property! - It is impossible because the claims of the official model not only doesn't match the available facts and observations but are in reverse sequence of what we are observing and know for the Sun.

In an attempt to cover up the nonsense of Big Bang theory, which dictates, that the stars must be formed from only available (in this time) primordial staff – Hydrogen, the mainstream science give us the model, where our Sun supposed to be just ball of hot gases, but the available facts show that this explanation is absolutely incorrect.

The model claims that the Sun is a hot ball of gases - mainly hydrogen. Nuclear fusion in the Sun's core producing the heat. Some invented "radiation pressure " of "massless" photons preventing the Sun from collapse. Also, some "mysterious" mechanism preventing the Sun to explode as one ordinary enormous hydrogen atomic bomb.

The current model of the Sun dynamics is based mainly on gravity and pressure. In this gravitational model, many properties and interactions are labelled as "Puzzles." The problem with the proposed model is there were our observations show a completely different picture and property of the proposed model. - The heat distribution of the Sun layers is in a reverse sequence of the proposed. One of the "Puzzles" is that the Sun has a surface and atmosphere! - Well known fact is that gases cannot have a surface and atmosphere above! The visible internal Sun temperature of the sunspots vortexes is 4,000K, followed by the Sun's photo-sphere 5,800K, and above it is the Sun's Corona with temperatures up to 2,000,000K.

The sequence of convectional heat exchange mechanisms of the proposed model is exactly in the opposite order! - The official Sun's Model cannot explain such opposite and bizarre heat sequences. Therefore, the creators of the gravitational, nuclear fusion Model of the Sun, declare again that this is a "Big Puzzle"!

The known fact also is that If the star's interior for some reason reaches the necessary conditions for nuclear fusion (how the model is claiming to be), nothing will stop the star exploding as one ordinary hydrogen bomb, but the next "Puzzle" (for them) is that the Sun is not exploding!

To solve these problems, we have to start with consideration of the physical state and functions of the Solar body. "They" claim that the Sun is just a hot ball of gases, mainly hydrogen, but the numbers don't add, because the Sun's specific gravity weight is 1.4 metric tons per cubic meter! - This is an impossible number for their model because the specific gravity weight of Hydrogen is 0.007L only! The liquefied hydrogen is 0,07L. Even the metallic liquid hydrogen is 0.7L. - All those three figures are far below the specific gravity weight of the Sun, which is 1.4L! The difference between the actual figures and the proposed figures is on a magnitude of 20 - which is equal to a 2,000% error! That's why I have started with the statement for the absurdity of the official Solar Model because even the specific gravity of metallic liquid hydrogen is 0.7, which is 700g/L. - This also is far below the Sun's specific weight of 1.4kg/L.

From physics, we know that no matter how much we are compressing gases, they always remain lighter than their liquefied state! From physics, we also know that we cannot compress liquids - their volume remains constant, no matter of pressure we apply. From those known and undeniable facts, it is more than obvious that the Sun cannot be just a ball of hot gases and even cannot be just liquid or solid hydrogen! - The Sun is much, much heavier than that! (They just avoid to mention these facts).

Under the enormous pressure, most of the interior of the Sun inevitably will form liquid metallic hydrogen with a solid metallic core, because it is obvious that the Sun's interior is not a million degrees hot, but is only 4,000K!

Opposite to the other gases, the hydrogen needs temperature over 1,000C to break the molecular bond and to transferred into liquid metallic form. This strange physical property of hydrogen also looks like to be carefully "chosen" to support the energy exchange of the stars – (similar case with carefully "chosen" physical property we have with water) - it expands when freezing! Scientists suggest that metallic liquid hydrogen even in this high temperature is acting as superfluid and superconductor! - This explaining how and why the Sun's interior is cooler than the surface, because of its heat superconductivity The enormous Sun's magnetic field also supporting this claim and is clear indications of fluid activity of the liquid metallic core. The ability of Sun's structure to absorb, transform, and emit energy is amazing! Such a structure will be presented in every star, and this structure obviously cannot collapse, how the "model" is claiming. Liquid and solid materials are not collapsing! The assumptions for collapsing of stars is just baseless fantasy, necessary to be created, to be in the line of the age of the Universe, proposed by Big Bang theory.

The understanding that a star's structure is formed not of gas, but is mainly liquid with a solid metallic center also rules out the officially promoted theories for the existence of Neutron stars and "gravitational collapse" - producing Supernovas! The reason for the exploding stars is absolutely different than the officially promoted, and we will consider it separately.

The conditions for thermonuclear fusion, according to the scientists, need temperatures above 10 million degrees but the sun's surface temperature is a miserable 5,800 degrees! How is it possible for the Sun's surface to be 5,800 degrees only if is sandwich between two layers millions of degrees hot?

The neutrinos coming from the Sun is not enough to explain nuclear fusion. - This is another indication that the current Sun's model is incorrect! But the guys are clever! They "discover" that the neutrinos coming from the sun are "enough" to explain nuclear fusion, because in the way from the Sun to Earth the neutrinos are "changing flavor"! Brilliant! Congratulations! They even discover that neutrinos have mass! Fantastic! - Everybody knows that there is no such thing as a mass-less particle because the particle has to be made from something! - And for this genius discovery "they" got a Nobel Prize! ("They" must be very, very intelligent). But... there is a problem...! "Small problem" - The neutrino detector hasn't been situated between the Sun and Earth, but underground! - How they manage to detect, measure, and decide that the coming neutrinos from the Sun are changing the flavor? In every square centimeter surface, every second is passing billions of neutrinos coming from space! In the neutrino detector, they are detecting only about <u>ten neutrinos a day</u>! What the statistical credibility of such result is? – obviously Zero!

The electromagnetism is a force that is 10^{38} stronger than gravity, but the mainstream scientists are stuck stubbornly to a gravitational model and are refusing even to consider the electromagnetic processes of the Sun and stars! For the strength of the Sun's electromagnetic field, you can judge form this fact:

The Sun gravity is enormous, and despite this, the electromagnetic corona mass ejection ejecting and accelerating billions of tons hot plasma in a matter of seconds up to ¼ of the speed of light into interstellar space!

The sun activity is on a regular base of the 22-year circle, where every 11 years are changing its polarity! - This cannot be explained with nuclear fusion, but with electromagnetic interaction based on a regular rotational cycle of the Sun, and electromagnetic energy exchange between the Sun's and the Galactic magnetic fields.

This is only a "small" example of the real pile of obvious miss-assumptions, and unexplainable phenomenon, which could be explained with electromagnetism, but not with gravity, pressure, and fantasy.

"They" assuming that Sun is just ball of hot plasma, but plasma is electrically charged particles and molecules that obey more electromagnetism than gravity and "They" ignoring even this fact and giving us the absurd scenario of gravity-driven Solar Model.

If we are looking at the real footage of the loops of plasma above the sunspots, we can see by ourselves that when they burst, the plasma streams are accelerating away from the Sun with continuous increasing speed! - This acceleration is opposite to the gravitational pull of the Sun and cannot be explained with gravity because it is an obvious sign of continuous electromagnetic acceleration against the Sun's gravity!

It is well documented the presence of substantial Sun and Galactic electromagnetic fields. Those two well-known electromagnetic fields – (Sun's and Galactic) have to interact at some point where this interaction inevitably will produce heat! This is exactly the phenomenon that we are observing in the Sun's Corona! Interaction and canceling of magnetic fields!

Here is an example of the absurd situation with the official model:

If I try to explain the function of the microwave oven and the arc welder with gravity and pressure interactions of matter, you will label me as stupid and crazy! Why? - Because it is obvious and well known that the heat of the microwave oven and the arc welder is produced by electromagnetism and gravity has nothing to do with it! An absolutely similar situation we have with the Sun, but in this case, everything is in reverse order of logic, knowledge, observational data, and facts! - The people who are explaining the plasma interaction of the Sun with electromagnetism are labeled as uneducated. Why? Who is the uneducated, the one who explains the interactions with facts and law of physics, or the one, who is ignoring all the facts and declares them a "Puzzle"?

The attempt to slide under the carpet those facts and proclaim them as a "Puzzle" is a very disturbing way to do science!

Supernova:

From all the evidence which we have considered above becomes obvious that the electromagnetic interactions are the prime source of the stellar energy. The star's core is mainly liquid metallic hydrogen, which superconductivity acting as a cooling substance of the star interior. This providing the stars with a very good radiation system, which acts as a cooling mechanism for the star's interior! The liquid star interior is not allowing the star to collapse and explode as a supernova. The reason for supernovas explosion is not a collapse, it is different. - In rare cases the electromagnetic interaction between the star and galactic magnetic fields becomes extreme. The heat production exceeds the star's ability to cool down. This leads to the situation, where the star's interior start's getting hotter and hotter and when the star's core reaches the required temperature for nuclear fusion - (about 10 million degrees), and then the star explodes as one enormous hydrogen bomb, which we are calling a 'Supernova.' - This is a simple physical process and is no mystery involved! There is nothing mysterious in Solar and stellar physical interactions if we are considering the facts with logic, with the laws of physics, and without preference, ignorance and hidden agenda.

After the explosion of the Supernovas, there remain nebulas. The composition of these nebulas without exceptions is about 90% hydrogen, 6% helium, and 3%-4% other elements. This is proving beyond doubt that the claim of the official model is incorrect, because in this model, the star exploding only, when all the hydrogen and helium is transformed in heavier elements, but the composition of the remains of the supernova explosion proving that this is not the case!

This proving that the energy of the stars is not a result of Nuclear Fusion, but is the cancellation of enormous electromagnetic fields!

Neutron star, Pulsar, and Magnetar

This subject is an extreme example of a totally wrong concept, unacceptable ignorance of physical properties and lack of common sense. It is an attempt to explain the most powerful electromagnetic field and electromagnetic emission in Universe with the most unlikely elementary particle, which has no any electromagnetic property or ability to produce any electromagnetic field – the neutrons!

According to the official theory, the Neutron star, Pulsar, and Magnetar can be considered as Neutron stars with slightly different magnetic properties. The difference between them is the strength of the magnetic field and the way how we can see them and their emission.

The story starts with a big controversy. In 1967 young student Jocelyn Bell has notice some rhythmic cosmic radio emission. For beginning her professor, Anthony Hewish was suspecting this to be a beacon of intelligent society (Little Green Man). Fred Hoyle and Thomas Gold come with the proposal for rapidly spinning neutron stars.

The controversy starts with awarding Nobel Price to the supervisor of Jocelyn Bell – (Hewish) for her discovery of Pulsars. The unethical act of the two people who accept the prise and the glory for the discovery of the young lady has steered the astronomical societies. Even Fred Hoyle argued that Jocelyn Bell at least should be included in the prize. Anyway, this is not an isolated incident with the wrong awarded Nobel Prise. Nobel Prize very often is used for political reasons or as a validating tool for some bizarre and weird theory as proven fact. A good example is when the Nobel Peace Prize was awarded to Obama, who started his presidency with two wars and ended with six. There is known a few Nobel Prizes for physics, aimed to prop up the impossible scenario of Big Bang theory - (as the superfast inflation), which is a brutal disregard of the law of Physics.

Here is a short explanation of the official version of neutron stars; their property and how they are supposed to forms:

At the end of the active life of a star, the internal radiation pressure of the star becomes weak, and this leads to sudden gravitational collapse of the star, which explodes as a supernova. In this violent event, the electrons and protons of the central core of the star are forced to fuse together and form neutrons. The enormous pressure of the explosion is pressing the neutrons tightly together to the extreme density of the atom nucleolus. As a result of this, the average diameter of the newly formed neutron star is about 12-20 km wit 1 to 3 solar masses. A cubic inch of its material will weigh on earth about a billion tons. The pulsars are spinning incredibly fast! The spinning rate is measured by milliseconds! The fastest measured spin is 1122 rotations per second, which are 67,320 rotations per second! The magnetic field of Pulsar and neutron star is quadrillions of times stronger than the earth's magnetic field. The Magnetar's magnetic field is another 1,000 times stronger than this! The energy outburst from neutron stars is enormous by any standard! The Magnetar SGR 1806-20 is emitting in one second the energy our sun producing in 1 million years! The source of this energy outburst supposes to be the kinetic energy of the spinning star and its interaction with the electromagnetic field (dynamo effect).- Fascinating story, isn't it?

Let see how credible is this "fascinating story," because as usual, we are given the most unlikely scenario for the events where the facts are showing something very, very different than the officially promoted story:

- The first fundamental misassumption is that the stars are a hot ball of gas and they are collapsing. As we mentioned above, the specific gravity of the Sun is 1.4, and this is ruling out the credibility of this claim.
- The Neutron star is a dead star! - How can dead stars produce a trillion times more energy than an active star for millions of years?
- How the electrically neutral particles of the dead star (neutrons) can produce a billion time's stronger electromagnetic field than an active star which is formed of 99% electrically active plasma and liquid metal?
- The life of free neutrons not bonded in the atomic nucleus by the strong nuclear force is only 15 minutes! How could the neutrons survive that long in the "neutron star" and not disintegrate?
- The kinetic energy of the neutron star supposes to be the "source" of its enormous continuous energy output! But the rotational kinetic energy of collapsing star is getting less! – Is not increasing! There is not known any mechanism of continuous self-creating kinetic energy! The law of physics is quite clear on this subject: - You cannot create energy! – In reality, "They" trying to convince us that the rotational kinetic energy of our Sun is trillions time more potent than the potential nuclear energy of its enormous hydrogen content. Are we should be so naïve to believe this?
- Sometimes the pulsating emission of pulsars is getting faster, some time is decreasing, which is contrary to any logic! But the official explanation is that the neutron star is absorbing matter. I am sorry, but this is impossible! The centrifugal force of such fast-spinning objects will not allow this to happen! A simple explanation as a discharge of capacitor on a grand scale is a much more credible explanation but is unacceptable for them and their gravitational model.
- The energy output of known neutron stars exceeds the mass-energy

limit allowed by the law of physics with 1,000 times! Why should we believe assumption which is known to be up to 100,000% out of spec?

Simple arithmetic and law of physics can tell us that if the radiation outburst of a neutron star is generated by the rotational kinetic energy this energy outburst will act as an effective rotational brake and the neutron star will l lose its rotational energy in a matter of days, or weeks! The fact that this is not happening is telling us that the energy source of the hypothetical "Neutron Star" has completely different origins!

To be easy to understand the impossibility of their model, I will give a simple example, which many ordinary people know - When you start a petrol generator, the engine run free and happy until you start using electricity. The more appliances you connect, the more the engine starts struggling because of the electricity consumption act as an effective brake on the spinning generator! How bizarre will be the claim, that when the generator runs out of fuel the rotational energy of the fly-will will be able to run the generator for millions of years and the electricity output will be a million times more, than when the generator running on fuel? - The same scenario is valid for the proposed neutron star – the enormous energy outburst must have a continuous energy supply, or it will act as a rotational brake and will stop the object spin in minutes!

Absurd models and explanations as this we can have, only, when the ethical standard of those scientists is in direct conflict with their knowledge! - To explain one of the most powerful electromagnetic discharges in Universe with the most electromagnetically impotent substance - neutrons, and to use the billions time weakest force of gravity - in order to ignore the existence of electromagnetic processes in Universe is a very disturbing way to explain the world and to do science!

Quasars

Quasars are the most energetic objects in the entire Universe. They have incredible strong radio emission and can outshine thousands of galaxies for an extended period of millions of years. All quasars are assumed to be billon of years away from us because they have significant light red-shift.

The official explanation of these objects is as usual based on gravity, mass, and ignoration of the vital evidence. The official explanation is that Quasars are Black Holes feeding of gas and plasma. Their mass estimation is up to a billion times more massive than our Sun. Well...well... but there are also some significant problems with this official explanation:

First, the existence of Black Hole is not proven yet, and the law of Physics is not allowed the formation of such kind of singularity! (Planck constant)

Second miss-assumption is the enormous assumed mass of the Quasars. - The observations show that the number of Quasars has been ejected from their host Galaxy. There is no known physical mechanism or force capable of such a "massive object" to be ejected from a diffused cloud of stars! This single fact is proving that Quasars cannot be such massive objects!

The third misassumption is the distance to the Quasars. Their light red-shift is mistaken for fast recession and huge distances, but the observations show that some quasars, which is assumed to be billions of years away are physically attached to the much closest to us galaxy and the assumed distances of the two connected objects are differentiating with billions of light years distance! - (Such "accuracy" is unacceptable)

What actually is the truth for these mysterious objects? The Quasars are simply "vortexes" of gas and plasma attracted by some powerful electromagnetic field and nucleus. The fast-spinning plasma in this magnetic field, generating extreme heat and produce enormous energy output. The incredible strong magnetic field of the Quasars is responsible for the electromagnetic light redshift, (this is the author's original assumption) - which mistakenly is assumed as recession and great distance to the Quasars. In order to explain the incredibly powerful energy output with gravity and to deny the electromagnetic origin of this phenomena, the mainstream science baselessly declare that the Quasars are incredibly massive!

Again, we are facing attempt something which is unknown, to be explained with something which "They" don't understand - (Black Hole), which don't exist!

The myth of understanding the atomic structure of matter

My admirations for the ancient philosophers, which without the luxury of the present scientific instruments has described with absolute precision the structure of atoms and matter using observations and logic. Those two vital components of gaining knowledge and understanding nature are missing form the present scientific community. A typical example is the atomic model presented to us from the top echelon of particle physics. As usual, in the structure of the atomic model, they are inserting mystical assumptions. As a result of this, the given model of the atomic nucleus becomes a bizarre assembly of hypothetical particles with invented properties, where the end result is shocking!

To explain the strong nuclear force, there are involved some sort of nuclear glue in the form of invented particles called "gluons." Despite the proposal of those hypothetical particles, this assumption doesn't explain the real property of atoms where was also added another hypothetical particle - "gravitons", and also some kind of "strange", "charming" and "colored" elementary particles but these inventions were not good enough and later even "God" was invited to help the "clever" guys, with the invention of the "God particle" - (Higgs boson). So far, all those bunches of invented particles cannot explain anything, because it is patch up job, where one by one, chaotically has been added all those incompatible particles and weird properties. Still, "They" are not providing any kind of an explanation of how the attraction mechanism working and how the "attractive particles" providing attraction. (The phenomenon of physical attraction I am explaining in a later chapter). I just would like to explain that the atomic nucleus is an energy cancellation center, which produces a strong attraction with short-range. The logical analysis of the proposed scenario will point to the fatal flaw of the present atomic model, where the strong nuclear force, responsible for holding the atomic nucleolus together supposed to be the property of the nuclear constituencies and is distributed evenly between the nucleolus particles. In this configuration, the number of the hypothetical gluons will increase in parallel with increasing the nuclear particles, and the atomic nucleus will be stable at any number of particles. That means that this scenario is allowing the atom to grow exponentially even to the size of the entire Universe. – (Well done!)

If the ancient philosophers have chance to consider this bizarre model, they will explain to the cream of our intelligent community with simple words, that the strong nuclear force cannot be distributed evenly between the nuclear particles, but is situated in the center of the nucleolus - that's why when the number of nuclear atomic particles is growing, the outer layer of particles start getting loosed with the increased numbers and the nucleolus become unstable above the 82 elements of the periodic table. Unfortunately, logical considerations of facts are not part of the present "scientific" practice!

STEPHEN HAWKING, BLACK HOLE EVAPORATION, INFORMATION PARADOX, AND THEIR LOGICAL SOLUTION

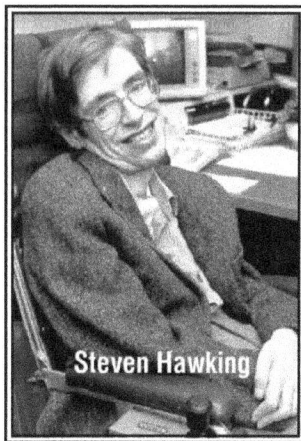

Stephen Hawking

This case is an extreme example of the way that science is manipulated, suppressed, and diverted towards a dead end. I have full respect and deep sympathy for the person of Stephen Hawking, and I believe that he is an intelligent person, but at the same time, I have complete disrespect for the agenda that he, or the group of people behind him, is promoting.

112

The difference between intelligent person and pseudo-scientists is this: the intelligent person uses his knowledge to make sense of the world and nature and adjusts his knowledge to the functions and properties of the physical world. The pseudo-scientists adjust the properties of the world to his equations and invent non-existing phenomena to justify his results. – This is exactly what we see in the present 'Western science.' This unethical non-scientific approach also is the basis of the Hawking black hole 'information paradox' and the weird 'evaporation' of Black Holes.

The existence of Black Holes in the assumed form is not proven! There could be massive objects in the form of normal matter because the law of physics not allowed the particles to be compressed further of the Planck length 10^{-35} m.

What actually is this 'hypothetical' paradox? – It is simple scientific arrogance, ignorance, or example of extreme stupidity.

As usual, when the pseudo scientists cannot explain something, they invented a mystery. This is the case also with the "virtual particles," which is in the center of this subject. Their existence is not proven, and is questionable because the law of physics stating: 'you cannot add or take away even single particle to the universe because you cannot create, or destroy energy"! But the pseudo scientists getting around the law of physics with invented mysteries and skillful mathematical manipulations.

Here are a little history and explanation of the phenomena: The world celebrates the genius of Stephen Hawking for his discovery of the "evaporation of Black Holes." What actually is this? – I am starting with the assumption that the universe is a closed physical system and we are not allowed to add or take away even one atom or particle from this system (It is supported by the laws of physics and explained in a later chapter). This is a basic fact, which Hawkins's theory for the Black Hole evaporation violates under cover of complicated calculations. This is a theory where the laws of physical fantasy and weird mathematics go against any logic and reason. - Hawkins's theory for evaporation of black holes uses 'virtual particles' and many controversial assumptions, such as the assumption that only one of the pair of entangled virtual particles(which pop out from nowhere) (or from the vacuum), is falling into the black hole but the other one is escaping. There is no physical reason for that to happen, and there is still no hard evidence that those virtual particles really exist! In reality, it is more likely that both particles will fall into the black hole and will increase its mass. Actually, there is no physical reason for the virtual particles to fall into the black hole because according to the 'theory,' they carry force, but no mass! So…, when there is no mass, there is no gravitational force and reaction to gravity! - Full stop! Obviously, there is no any reason those virtual particles to fall into the black hole!

The next wire assumption is that the Black Hole is emitting gravitational force I am sorry, but this is not my statement – (is theirs) that gravitational force does not exist and is just bent space! – shamelessly they are using conflicting statements and conflicting properties when they need it.

The third weird physical reason is that Hawking assumes that one of the virtual particles will have negative mass! (Or constructed from 'negative energy,' perhaps)? So far there is no any sign or any proof that such a thing as negative energy or negative mass can exist and such assumptions for particles with negative mass and negative energy are absolutely unacceptable! In mathematics, you can have negative values, but in physics is no such thing as negative mass and negative energy! We know of the existence of antimatter, but antimatter carries the same amount of positive mass and positive energy - the antimatter just has reversed internal electrical polarity and forces. In contact with ordinary matter, it is annihilating and releasing a huge amount of normal positive energy!

In reality, even the existence of the virtual particles is unproven and questionable. The existence of the Black Holes in the proposed form is not proven, and violating the law of physics! You cannot construct a scientific theory of the basis of bunch unproven and non-existing phenomena. Black Hole has no boundary! The event horizon is an imaginable optical illusion for a hypothetical observer, like the shiny and shadow part of a picture. The event horizon has no any different physical property than any other part of space! What will be the physical mechanism that the quantum information choosing to "stick" exactly to this chosen by Hawking region with no any physical difference?

All those incorrect assumptions that Hawking uses to construct the base of his theory make it scientifically absolutely unacceptable. It is a more mystical fantasy than actual science!

The Black Hole Information Paradox

The so-called "information paradox" actually comes from the desire of the establishment to cover up the real properties of quantum information. Quantum mechanics states that quantum information cannot be created, cannot be copied, cannot be destroyed and cannot be lost. The pseudo-scientists created a hypothetical non-existing problem with the information link between entangled quantum particles in the hypothetical scenario, where one of them falls into a black hole, but the other entangled particle escapes. Hawking and his group create an absolutely unnecessary so-called "paradox" to spread further confusion about the amazing properties of the quantum information. In reality, there is no paradox at all, because quantum information is not material and is not part of the physical property of matter! –It is information, it is the part of the 'software' of the universe, and it is the invisible 'hand' of the law of physics!

(The detail explanation of the origin and property of quantum information is in a later chapter).

The gravity of black holes does not affect quantum information because it is not material! Quantum information operates freely in the whole universe and has no physical barrier or condition to stop or restrict it –as temperature, pressure, gravity, distance or time, and there is no any information paradox at all!

Black holes, (if they really exist in the form 'they' are proposed), must be the ultimate milling machines of the universe. Any form of matter falling in will disintegrate into its basic form of energy and will be ready to be re-injected and re-used back in the universe. The fact that everything in the universe is logically built and perfectly balanced leads to the conclusion that there must also be incorporated a simple system for recycling of matter back to its prime form - energy. The so-called final destiny of the matter – black holes which transfer matter in its basic form, energy – really must have a mechanism for releasing back this energy, and Hawkins's proposal looks absolutely impossible scenario.

Here are more details: Group of educated pseudo-scientists start fantasizing and doing mental exercises with black holes. They have stumbled on the invented 'problem' - (the information paradox) - that when some particles fall into a black hole, the particles take inside the information which "they possess."

In an attempt to save the principle of quantum mechanics, that the information cannot be lost, they declare that this is a 'big puzzle,' or 'mystery' because the fallen-in particle cannot share information with its entangled partner who has not fallen in the black hole and is outside. There has been a very heated debate (a scientific war) and arguments between the two groups of intellectually impotent people. - (Similar scenario to the middle century debate– 'how many devils can stand on top of a pin'). One of the group insisted that the fallen information in the black hole is destroyed; the others insist that the information cannot be lost or destroyed. In a stroke of 'genius', one-day Stephen Hawking put some kind of 'contact glue' on the (physically non-existing) 'surface' of the black hole called - (the event horizon) and declared that the particles fall in, but the information sticks to the surface of the (non-existing) - event horizon and is emitted back by 'the Hawking Radiation'. (He even makes a $100 bets with his opponent Preskill on this!) The event horizon is only an imaginary boundary, and in reality, there is no any physical difference with the other regions of space. There is nothing to stop and preserve the information, rather the stupid assumptions. Every average intelligent person with a minimum understanding of physics will see the flaws in the official version of this mental exercise called the 'Stephen Hawking Black Hole Evaporation.' **He explains something that he doesn't know with something he doesn't understand to solve a phenomenon which doesn't exist!**

What do we have, actually? (If the black holes do exist in this form) We have a black hole, where the matter or particles fall in. The gravitation of the black hole is supposed to be enormous; nothing material can escape, not even light!

OK, but what is the property of the information? Is the information material? Does the information have mass? - No! The information is not a material substance and cannot be affected by gravity! There is no any information, which can be affected by gravity! - So, what is the problem the arrogant ones talking about? How it is possible that the gravity of the black hole will affect something that is not material? Does gravity affect music and is music falling on the ground? Is the gravity affecting the story, the computer program, or our emotions and fillings? What paradox the corrupted is talking about? The reality is simple and obvious –quantum information is not material, and gravity has no effect on it! The information can travel freely in and out of the black hole and will do not care about gravity and stupid assumptions!

The other 'genius' invention of Stephen Hawking is the mechanism for the evaporation of black holes. He assumes that some mysterious entangled particles are popping out from the vacuum and they must obey his will and for an unknown physical reason only one of them has to fall into a black hole, but the partner must escape. And because the falling particle has "negative energy," this supposes to deplete the mass of the black hole. – Wonderful, isn't it? He didn't realize, that the matter in the black hole is the same as the matter outside, and if the virtual particles is canceling the existence of the matter inside the black hole, in same way the virtual particles should canceling the existence of the matter outside of Black Holes, and effectively the universe will disappear! The fact that we are not observing such a scenario is proving beyond doubt the nonsense of the Black Hole evaporation theory! Wire hypothesizes, but even the existence of a black hole is not proven yet! And despite this fact, the Nobel Prize for Physics for 2017 has been awarded for the recording of gravitational waves of merging Black Holes! – Fantastic!

Hypothetical image of a black hole, which existence is not proved yet!

How I stated above, the existence of any negative energy or negative form of matter is unknown to physics! – Simply do not exist! And why only one of the pair must fall in the black hole and not both of them is not clear, but this was declared to be one of the greatest achievements of science and triumphs of the human intelligence!

I am sorry, but if this nonsense is a triumph of our intelligence, then what do we have to call stupidity?

I have started this chapter with examples of how intelligent people solve problems - with logic and knowledge. And here we are facing exactly the opposite approach!

One of the problems of the 'pseudo-scientific society' is the evaporation of black holes. The 'smart guys' state that black holes are the final destination of matter, but observations of the universe show that this is not the case! So... the pseudo-scientists start fantasizing and inventing their own mechanism for the evaporation of black holes. The evaporation of their total connection with common sense and reality is stripping them from the ability to see even simple and obvious natural processes: I will start with the property of gravitation force. From the famous formula of Einstein, $E=mc^2$, we know that matter is a concentrated form of energy. In the same way, we can consider that the gravitation force is (energy), which is mass.

I will give you an example. - Physicists calculate and transferred the gravitation force of the hypothetical dark energy (negative gravity) into mass and state that this is actually 74% of all mass of the universe. - Any form of force or energy is actual mass or a form of mass! When matter radiates energy (gravity) it actually loses mass! They say that gravity is not a force, but a bending of the space. OK, even in this scenario, the massive objects need to exert force to bend space, and this physical interaction between mass and space can be only at the expense of the mass of the object; one way or another will be the same result! I would like to make clear this point because this is vital for understanding how black holes (if they exist in the proposed form) how are really evaporating:

When we cut out all the nonsense, we know that black holes supposed to emit extremely powerful gravitational force. As I stated above, gravity is an actual mass! And if black holes emit extremely powerful gravitation force, this in practice will be the emission of mass! The emission of gravitational force will be at the expense of the internal mass of the black hole. And continuous gravitation emissions slowly will reduce and evaporate the hypothetical black hole! - It may take a lot of time, but the universe is not rushing; it will be here to stay long after us. It is a simple natural physical process -part of the universal recycling mechanism, where the mass of the black hole comes back out in the form of the primordial state of matter – 'energy,' ready to form new structures again! There is no need to be involved in any mystery or weird assumptions, complicated calculations, or puzzles! - This is a simple example of how two types of people solving scientific problems: one with facts and logic, the others with pure fantasy mysteries and ignorance.

There are many different opinions and theories about global warming, but the simple truth is that we are extracting from the earth and pumping out enormous amounts of carbon dioxide, and only a stupid or crooked person can advocate that this will not affect the global balance of nature. Most of the politicians brand themselves as champions of the battle against global warming.

Let us consider the facts of what actually stands behind this? – Because this is vital for us, for our children and for the future of our planet!

One of the most vocal politicians involved in the global-warming issue is the former US Presidential candidate, Al Gore. He claimed that his concern was for our future, and he had a solution for reducing global warming. What actually was his solution? - To make carbon dioxide, the product of trade on a global scale, where he and his friends would determine how much each individual company in the world would have the right to produce carbon dioxide. He and his friends had already established carbon-accessing companies and were pushing for their system to be implemented by the United Nations. What, effectively, will this proposed system do? - This system will open the door for political pressure to be applied to each country, and each company in the world; it will open the door for the corrupt individuals to impose taxes on every single company in the world. Those individuals will be the middle-men in each trade of carbon emission, taking commissions. Those individuals will have the power to dictate the terms of the development of national industries. Those individuals will have the power to impose penalties, fines, and the closure of each company they don't like, or which compete with their own. Such a system is an open door for bribery and corruption on a global scale. - This is the vocal idea of people like Al Gore. The former prime minister of Australia Tony Abbot, when his successor imposed a carbon tax, made a very good statement: 'The carbon tax will not clean the air; it will clean your pocket only.'

I cannot comment on the legislation of each country, but I will give you an example from the country where I am living - Australia. I had the idea to install solar panels on the roof of my house. I am qualified to do this job by myself. I have extensive knowledge of electricity, building constructions, engineering, and sustainability. I approached the institution to find the requirements and regulations. What I found was shocking! With all my recognized qualifications, I was not allowed to install my solar panels, even when I asked for a qualified inspector to certify the installation. I was told that I must use the government-recognized companies; I must buy the products from them, and use the government's financial incentives.

What are the actual details of so-called government-subsided solar energy? The details are shocking: If I bought good German solar panels and good quality regulators, I would have spent about $4,000.

If I used the government-recognized company, they would have provided me with Chinese products and the system would cost about $18,000. The government incentive at the time of my inquiry was about $8,000. So, I would be $10,000 out of pocket for a system that costs $4,000. Taxpayers also would be $8,000 out of pocket. The profit would go to the electricity company which would not offset my electricity production from my consumption. - When the solar panels start producing daytime electricity, the electricity supply company would buy my electricity and pay me 4 to 6 cents per KWH. But in the evening, they would sell me back my own electricity for up to 36 cents. On top of this, they would charge me distribution charges and line maintenance. Effectively, the government and the legislation would force me to pay the electricity distributors about 30 cents for every kWh that I am producing. - This is the bitter reality of the government's effective destructive environment policy. The government claims that they are subsiding renewable energy, but this claim is only a public show, a pure scam that rips off ordinary people for the benefit of companies selected by the government.

Australia is a hot country, and significant amounts of energy are wasted in cooling houses. If the government was really dedicated to reducing global warming, why are such simple measures are not regulated in the building permits and building codes, like the following two?

Heat mostly comes from the roof. Legislation should disallow a dark roof color. On sunny days the temperature in the roof cavity can reach over 100°C. If the exhaust flow of the evaporated air conditioner is directed through the manhole into the roof cavity, this would cool the roof cavity and will stop the heat coming from the roof in the house! Only those two simple measures alone would cut electricity consumption by 30%.

The other myth officially promoted is that if we plant more trees, this will reduce the amount of carbon dioxide in the atmosphere. Yes, we need trees, but this is only partially correct. It works for a limited time only, and will not make any difference in the long term. In the present natural conditions which exist on earth, every planted tree has a destiny sooner or later to decay or to be burned. In both cases, the absorbed carbon dioxide will be released back in the atmosphere. I am not advocating not to plant trees; I just want to make clear that this is not a solution at all in the battle against global warming! The solution is not to extract carbon from the earth's crust and release it into the atmosphere! Back in past geological eras, the conditions on the earth were different, allowing dead trees on a mass scale to be buried and form coal. This is the difference – the extracted carbon from the earth's crust to go back into the earth's crust in a stable form such as minerals, not only as liquefied gas! The best solution is not to extract and burn carbohydrates. To be able to do this, we need honest governments, which will not do everything possible to ruin our efforts. I will provide a simple calculation to reveal the absurd stand of our government. There are about 5 million Australian households. If for each house were allocated solar panel to cover its total consumption (in the desert), where there are free land and more sunny days than the coastline, this would cost each household about $3,000. The total sum for every Australian to have free solar power and zero carbon emission is about $15 billion. The Gillard government was collecting $17 billion a year in carbon tax only! - This calculation explains the situation and what exactly governments are doing. These facts don't need any further comments.

The Sun and nature are providing us with unlimited energy. The governments are using any excuse not to use it because it is free! Their policy is that we cannot have anything for free, because they cannot tax us and make money! And this is the reason for the brutal, senseless destruction of our planet! – The money!

The truth about the heavily promoted electric car is exactly the opposite! Currently, they are the dirtiest cars on the road because they are heavy, very energy-inefficient because they use inefficient multiplied electricity transfers, and use electric power produced by burning coal and natural gas. That's why they are promoted and subsidized by governments. The claims that solar power is more expensive are false! You can judge by self: which is more expensive, to extract oil gas or coal, send it halfway around the earth by ship, refine it, and send to the customer who burns it, or the Saharan nations to send free solar energy to all Europe by cable? The claim, that there is no practical way to store the excess of solar power, is totally false! The governments and the pseudo-scientists are promoting bizarre and absolute impractical ideas for storing solar power to be used at night time. They suggest using batteries or molten salt reservoirs, or to produce hydrogen and burn it. The money for research in this area is deliberately directed towards an impractical, or dead end. What do I mean? - The research for nuclear fusion is a wild goose chase. Nuclear fusion is a hypothetical assumption! It is not proven to be correct by any experiment! - (In the previous chapter I have revealed the absurd temperature sequence of our sun). The other research is directed toward breaking methane into carbon and hydrogen, and then to burn hydrogen as fuel. It is a wonderful idea; only one problem! – It cannot be done, because the law of physics states that to break the chemical bond of a substance, you need more energy than when you bring those substances together. For the same reason, the patent office does not accept and consider applications for engines without fuel, or engines that produce free energy from nothing. It is against the laws of physics! - Simply you cannot produce energy! Full stop!

The truth is that there is a very elegant, efficient, and cheap way to store the daily solar power excess. This is the system of two lakes. One of the lakes is higher than the other. In the daytime, the solar power excess pumping the water from the lower lake into the higher lake, and at night time the returning water from the higher lake turns the turbines and produces electricity. This also can be done with one lake on a higher sea coastline. Day time, pump up the water from the sea; night time, return it back through the turbines and have solar power energy for the night. The efficiency of this system is up to 95% of the produced power because the electric pumps have an efficiency of 96% and water turbines 99%. The running cost of this system is minimal, compare to all other systems, which need an army of workers, expensive maintenance, and regular battery and equipment replacement.   <u>This is a cheap, efficient, and widely available system! It is solving the energy demand and global warming!</u> - That's why this system is carefully hidden from the public! The biggest danger for the planet is the limit of the tipping point of the climate. What is it? - This is the point of no return! - In the permanent frozen Arctic soil and continental shelf are stored enormous amounts of methane. Methane is 20 times a more potent global warming gas than carbon dioxide. If the tipping point is reached, the melting and releasing of methane on a grand scale will start, and nothing will be able to stop it. Then after a while, when the oceans get warmer, another wave of methane from the frozen methane layer in the seabed will finish the terrible job. The geological record suggests that in the past there have been periods with average global temperatures higher than the present by 15 to 20 degrees! In such temperatures, humanity cannot survive. Nothing can survive rather than some bacteria! The crops will fail. Nothing can grow in such conditions if daily temperatures reach 60 to 70 degrees. The weather will be too extreme. Oceans will lose the oxygen, become acidic, and marine life will be extinct. The only possible place where only a small number of people could survive for a while in the Arctic, but the lack of food could return them to cannibalism before they perish!

But this not all the story, this is just the immediate effect of our activity, which will bring the second wave of global warming! – The real warming, with potential to reach $330C^0$, which is almost 100 degrees higher than the melting point of tin.

The second wave of global warming:

At the moment, we have countless climate "specialists." Some of them is measuring the air composition, others studying the glacier ice samples, other volcanic activity, and on...on... and on...!

The politicians are picking up the scientists of their choice – that means, that they are picking data and is acting in the interests of the biggest corporations no matter of the real consequences for us.

We need a little bit more intelligent approach to be able to understand what really standing behind the Earth's climate balance. We have to start our consideration on a grand scale, from the top down, not to use the chaotic pile of climate data collection to build conflicting wire hypothesizes and unrealistic climate models.

What are the real facts telling us?

The position of Earth in the Solar system is the major factor for the temperature and conditions on our planet.

The Earth orbit is between Venus and Mars. Venus surface temperature is 460^0 C, Earth surface temperature is 16^0 C, and Mars is varying significantly, due to lack of dense atmosphere, but is between -50^0 C and $+20^0$ C. – (surprisingly, It is very close to Earth's temperature). The amount of Sun energy those three planets receiving is: Venus $2,500 wm^2$ / Earth is $1,300 wm^2$ and Mars receiving $650 wm^2$. (It is assumed that in the past geological eras the Sun activity has been lower!). From those figures, if we consider that Earth receiving 100% Sun energy, relatively Venus having 70% more energy, and Mars receiving about 45% less than Earth. But there is something else, which is significant! Venus is reflecting back in space about 70% of the Sun energy when Earth is reflecting back only 30% of the Sunlights, which effectively doubles the amount of the received Sun energy by the Earth's surface compared to Venus! According to those figures, the Earth temperature should be scorching $330 C^0$ -(100^0 above the melting point of Tin!). But in reality, our surface temperature is $16 C^{0,}$ which is 300^0 C less than what should be! - Why? ... Why is such a difference? What is cooling our planet at such a significant rate? - To find the answer, we have to look and consider the major contributing factors of temperature balance! The most powerful cooling mechanism of Earth I believe is the cloud cover. Opposite of the claim of the scientific community, which believe that the clouds are trapping the heat, I believe, that they are doing the exact opposing job and cooling the planet. The logical conclusion for this we can get if we not concentrate on the local effect of cloud cover, but to consider the Earth as one hull system. Then the white color of clouds obvious will reflect back in space most of the incoming heat from the Sun. The white color, compared to the dark blue ocean water, make a huge difference!

From the paleontological record, we know that in one stage Earth even been complete frozen – as "snowball" due to the increased amount of oxygen and lack of CO_2. It is very easy to figure out that Earth's reflective surface and the atmosphere composition doing a very good job to cool down our planet and to sustain the surface temperature in the most comfortable zone for biological life. But we also should realize that the composition of the Earth's atmosphere is dependent on the health of the biological life of our planet. With this understanding, we can start using the data and smallest details to complete the picture of the Earth temperature and climate balance and see who is correct – the proponents, or opponents of global warming.

From the facts provided, it becomes obvious that the composition of our atmosphere and surface reflectivity is a major factor, which protects us from catastrophic temperatures. Events as volcanic eruptions and asteroid impact can distort the ecological balance of the planet temporarily, but the Earth's eco-system is recovering well when the conditions become right again and bringing the atmosphere and planet temperature back to the required level. This is a crucial understanding of the core of the phenomena! – That the key behind the significantly cool surface temperature of Earth actually is the biological life! With the knowledge of how high in reality the Earth's surface temperature should be if is no biological life on the planet, we can understand what in reality we are doing! We have to understand that the bacterial life, plankton, and vegetations are the real guardians of the planet because they are absorbing the sun heat and transfer it into biomass.

So… let consider the major factors which are contributing to the global surface temperature:

The atmosphere:

The key elements are its thickness and composition. The thickness of the Earth atmosphere is much thinner than the Venus atmosphere, but the thickness by self is not the major factor, because even a tick but the transparent atmosphere will not have a significant warming effect. The atmospheric composition, planet surface reflection, and cloud coverage are the major warming factors. Let see how those factors will affect the Earth's surface temperature in the near future.

In the conditions of continuous increasing of CO_2 and steady rising of the average global temperature, there is a so-called "tipping point." The tipping point of climate change is when the permafrost of the arctic start melting and releasing the methane trapped in the frozen soil. The methane is at least 20 times more potent warming gas than CO_2. The released methane will produce a cascading effect and will warm the oceans too. In the continental shelf is frozen the enormous amount of methane, which will be additionally released. This huge new volume of methane will start oxidizing and will reduce the amount of oxygen in the water and atmosphere. Reducing the oxygen level is a huge warming factor by self because the oxygen is cooling gas, but also its lack will reduce the ozone layer, which is protecting the biological life, and is absorbing a huge amount of solar heat.

The next factor is the cloud cover and surface reflection:

To understand better the difference which I am talking about, the reader can check it by self: on a hot summer day, put the thermometer on the bare soil, or sand, block the sunlight in order to measure only the surface temperature. Then repeat the same and measure the temperature of grass. – The temperature difference will be about 20^0 C. Then touch the roof of the white car and on a black car on the same sunny hot summer day. - This will give you an indication what is the difference for the reflective planet surface of the ice caps and the dark open water of the Arctic sea and the difference between forest, or grassland and lifeless desert! When you take into account the vast area of the global surface, and you apply these differences, then you will understand how high the global temperature could rise if we lose the ice caps and the green Earth cover!

The continuous rising of the global temperature soon will melt the snowcaps, which reflecting back about 90% of the sunlight and heat. On top of this, the increased temperatures will increase the water evaporation and will increase the cloud cover and even the thickness of the Earth's atmosphere. - Everybody knows how significant effect is the cloud cover - when you are camping, it makes about 10^0 C difference overnight, a dependent is the sky clear or is cloudy!

When we consider all those factors honestly, without any reservations or speculations, the picture is grim! Only the comparable thickness of our atmosphere working in our favor and probably will reduce the potential Earth surface temperature by a factor of 5, but all other major factors are still on the table and are definitely against us! There is not much comfort in a scenario were instead to reach the full potential of warming, which is about 330^0 C we will make it in the range of $60_0 - 70_0$ C. The difference is very similar to the scenario, where the sentenced person is allowed to choose to be shot or be hanged! - Currently, we are enjoying 16^0 C average global temperature, and if we are increasing it with $10^0 - 20^0$ "only", this will spell the end of the life of all complex cell organisms on Earth. – (All plants and animals)

The dumb pseudo scientists, which are measuring the content of CO_2 and making wire hypothesizes and predictions don't understand anything, where the real problem of our existence is, and how catastrophic could be the consequences of our destructive behavior! We have to realize, that each plant, each shoot of grass, each leaf is absorbing the Sun energy – (the heat) and by the photosynthesis, <u>they are converting the sun heat into bio-mass</u>. On top of this, each plant is evaporating water and cooling additionally the environment. The so-called "tipping point" of climate will produce acidizing of the oceans and acidic rain, which will wipe out the forests and micro-fauna of our planet. The ocean will stop absorbing CO_2 - An exactly this will be the real catastrophe - To lose the Earth's protective eco-system and its balance! (Not the air composition only). If we do this, nothing will stand between the scorching Sun and us! We are far too close to the Sun, and the possibility of rising the average Earth surface temperatures is in the range of 50^0 C to 100^0 C- is not how currently the politicians and their puppets talking about just a few degrees difference! Is time to face reality! - We have to realize that Mars is 70% further from the Sun. Mars has no atmospheric warming blanket, no green forest, no ice cups, and Mars has nearly the same temperature as Earth If we give Mars our atmospheric blanket with more CO_2 and bring Mars to the Earth orbit, what will happen? - Mars will become hot as hell, and this is exactly where we are going!- Unfortunately, we are much, much close to Venus...and Venus is real hell!

The latest observational data reviling <u>that only in our galaxy is 10 million trillion planets</u>! - The galaxy should be teaming with life, and have millions of advanced civilizations, but space is absolutely silent! Why? Are we really stupid enough not to realize what we are doing? We have to realize that the continuation of our existence is very unlikely and is not guaranteed at all! The Universe is not a free lunch! <u>We have to make everything possible to make sure that we will survive, not exactly the opposite!</u> Why are we destroying our life guardian eco-system? There is nowhere else to go, nowhere else to hide! There is no hope for us if we lose this fantastic, beautiful and amazing life guardian system, given to us from mother Earth! The Earth has the ability to recover, and will recover one day, and will produce beautiful life, and again will be the bright blue marble, - "the pearl of Universe," which has been given to us, and which we recklessly destroyed. The Earth will continue its journey in the future, in one better future, without the destructive parasites, which have called themselves - human beings!

If we have a correct understanding of the world, we will know, that Earth and the Universe have all the quality to be conscious of being on a colossal scale!

From this understanding, we should know that Earth can give life, and also can take life away. We have to understand that there is a limit on how far we can go, and if we cross the line, we have to face forces mush more powerful, that we can ever imagine!

I am sad to make this statement, but this is necessary to be known: If humanity continues to deceive itself and not come in real terms with the correct understanding of the World, humanity has no chance to survive. We have a very distorted perception of the world around us:

Everything we know for the world is carefully falsified, and as a result, we wrongly believe that space and time are incorporated. We believe that space is bent, we believing that the Universe is expanding, we believe, that all elementary particles are mass-less, we don't know what gravity is, what is electromagnetism, we don't know even what is the composition of our Sun, and how the Sun is producing its energy!

Distortion perception of the World and reality is a mortal danger for every living organism, any individual, any institution, and any civilized society!

The increasing gap and disproportion between our technological advances and our understanding of the world is a very dangerous prospect.

With great sadness, I can say that without a correct understanding of the World, humanity cannot make rational decisions and cannot survive.

We are branding ourselves as intelligent creatures, but we are acting in such a stupid and irrational way and creating endless unnecessary problems for ourselves. There are many elegant ways of dominance and make a profit, then the chosen corrupted way of endless wars, killings, and brutal exploitations.

Our distorted perception of the World making us believe, that we can do anything to the Earth and get away with this! – Unfortunately, the reality is much, much different than the officially presented model, and the consequences of Global Warming will be catastrophic!

This is not a scary scenario or a script for a movie, but an absolutely real and present danger as a consequence of uncontrolled greed and stupidity of our leaders. Our problem is not global warming; our problem is the corrupted system that does not allows the problem to be eliminated quickly and efficiently.

We need to stop this! Can we?

MEDICAL ETHICS, CANCER RESEARCH AND TREATMENT

The funds for cancer research are mostly directed towards the private sector. The fundamental flaw in this activity is the commercializing of people's health and turning the health system into one for profit. The pharmaceutical industry became very powerful and forcing governments and the health system to adopt rules, legislations, and procedures which would ensure maximum profit for those companies. They managed to suppress natural medicines which have been proven in many cases to produce the best results. Despite the billions of government funds and private donations, research for many unprofitable sicknesses is neglected. The pharmaceutical companies are aiming to develop drugs only for sicknesses of massive scale where the number of drugs sold would bring them huge profits. For hundreds of tropical diseases, there is effectively no research or cure. But the most disturbing facts are coming from the so-called "battle against cancer".

The number of affected people in the developed world is becoming alarming.

Is this accidental, or does something very wrong and disturbing lie behind this epidemic? Why do the so-called third-world countries not suffer the same rate of cancer as the developed nations? By any logic, the situation should be the opposite, because the developed nations can afford to eat the best, healthiest and most regulation protected food, and the living standards and conditions are much better than poorer countries. Yet, despite this, every fifth person in the developed nations will be touched by this terrible disease. Can this be purely coincidental?

I will go back in time. In the 1960s, there was a study among the scientific community about the cure for cancer. The optimists predicted that the cure for cancer would be found in the next 20 years. The pessimists assumed a time scale up to the year 2000. What happened, that we got it so wrong with the development of the cure for this nightmare on humanity? Does this situation not look suspiciously like the situation with physics, where research is directed deliberately toward a dead end? The banks state that opportunity makes the thief! Why would we put ourselves in a situation where finding a cure for cancer would stop huge funds for research and the institutions involved would lose billions from treating long-time terminally sick people with very expensive drugs and procedures? I cannot provide proof for this assumption, but the reader is able to put two and two together to find the answer for himself because the financial incentives are against finding a cure. - It is simple and logical!

My experience with the way how diagnosed patients are treated comes from two members of my family who have suffered cancer. We know that when cancer cells are located in a gland, there is a good chance if the gland is taken out intact and immediate treatment applied, then the patient will have a very good prospect of being cured of cancer. The last thing that the patient needs is the gland to be cut open and the cancer cells to be spread over all the body. The disturbing fact in both cases was that a biopsy from the swollen glands was positive, but the treatment was delayed unreasonably for 2 to 3 months. My questions and protests were received with a smile from the specialist's doctors and with a blunt assurance that the biopsy was not spreading cancer cells in the body. I am sorry, but this claim is absolutely wrong! Any educated person knows that compromising the integrity of the gland inevitably will lead to leaking of blood and fluids and cancer cells in the body. If you are fighting for your life, every single day could make the difference between life and death. After the invasive biopsy, your doctor should make everything possible to administrate the necessary cure as soon as possible. Instead of this, the doctor takes a one-month holiday overseas without informing you, and you have to wait one more month to see him after his return. - That is what happened with the two members of my family. – They were forced to wait for 2 to 3 months after the biopsy to receive treatment. Why? - Obviously to make sure that they will get sick and will need treatment for the next 5 – 10 years!

I will tell you one sad story, which I don't want to comment. The reader is capable to make its own conclusion:

In the time when the world was stunned from the resilient of "Mad Cao disease" protein, which remains potent even when the infected meat is cooked or baked. No known disinfectant is able to sterilize this terrible disease. The only option was to burn and berry the infected and suspected animals. This was done and many millions of animals were slathered and burn in England and other nations. At this time, a friend of mine - young intelligent dentists - George Nikolov tell me that he has the idea to make the surgical instruments capable to be dismantled and properly cleaned after each surgery because in the hinges of the instruments remain blood which is inaccessible for cleaning. This remaining blood in the hinges of the surgical instrument inevitably will transmit diseases from patient to patient. Dr. Nikolov first has tried to publish his warning to the world, but his article was rejected from all medical Journals. Then he receives a treat from the medical association not to publish this article in any other magazine, because this will create panic and many businesses will lose money. They warn him, that they will suspend his license if he does not comply with their demand.

Dr. Nikolov continues his work and successfully has designed a range of safe and elegant ways of dismantling surgical instruments in order to be properly cleaned and disinfected. He makes a patent application in the USA - US5618308A - 2004. The following action of the medical association was to inform Dr. Nikolov not to produce these instruments because they will not be approved for use in western countries. Despite this warning, the father of DR Nikolov - Dr. Nikolov senior (who also was a doctor) went back in his homeland - Bulgaria with a mission to produce these instruments there. Not long after his departure, we have received news, that he - Dr. Nikolov senior was brutally shot in his car. After this "accident" Dr. Nikolov junior decides not to go ahead with his invention and he still is alive, healthy and is allowed to work. His invention with the potential to save millions was suppressed in aim some to make more money from our sickness.

My kitchen scissors have a better design for cleaning than the "modern" surgical instruments!

We have created a situation where doctors and the pharmaceutical industry have no interest in curing you completely. If they manage to make you sick in the first place, then over the next 5 or 10 years you will need endless chemo, radiotherapy, scans, operations, consultations, and expensive medications for years. Those institutions will make a lot of money from you. My question is: are they stupid to cure you completely? - Obviously not! We cannot blame them because we created this system. They are just making the best of it. We created a social system where is no room for ethics and moral. The greed and profit in our social system are more important than anything else. The money becomes more important than our health, our life, and our planet! This system is our creation, and it is our fault because we tolerate it! Donations for cancer research will not make any difference! Our stand, dedication, and contributions for democracy and ethical standards will make the real difference not only to the battle against cancer but in all aspects of our life and future.

WHY IS THE UNIVERSE SILENT?

For nearly 50 years, humans have been searching for signs of life or other civilizations, but the universe is silent. What happened, are we alone? - It is possible, but it is not very likely. So... where are the others?
To get the answer to this question, we have to assess the situation on a grand scale:
There are two stages of civilization development - primitive and advanced.
I am sorry to tell you that we are in the primitive stage of civilization. I am even not sure are we can be qualified even to be called civilized, because from an independent point of view civilized creatures have to possess minimum social and moral values to be qualified as civilized, and they are:

- To respect the right of every living creature to exist.
- Not to kill other creatures and eat them.
- Not to destroy the environment and nature.
- Not to make war and kill its own kind.
- To poses ethical and moral values: trustworthiness, tolerance, loyalty, and respect to others.

From the list of minimum ethical, moral, and social values, we do not possess even one of them. So... the logical question is:
What will be the reason for other civilizations to contact us? What is our ethical standard, and what can they expect from us? If we do terrible things to ourselves, our planet and our creatures, what will be the guarantee that we will not do those things to them? - They will not give us a chance!

130

There is a public dilemma: do we have to announce our presence to other civilizations, and it is safe? I am sorry to tell you the truth. - We have already announced very loudly to the entire universe who we are, what we do, and what is our moral standard. Most of the people forget that nuclear explosions are sending very powerful and distinctive radio signals into space. Any intelligent civilization will pick them up easily and will know exactly what it is. Both the USSR and the US have tested and exploded thousands of atomic bombs in their military polygons. By detecting those nuclear explosions, advanced civilizations will know that indiscriminate killing is our real ethical standard. We don't have to send them messages with assurances of how 'humane' we are! They know this already. Even the author of the song 'Imagine' - (John Lennon), which we sent into space with *Voyager*, was violently killed. - Even this single fact can describe perfectly our real moral and ethical standard.

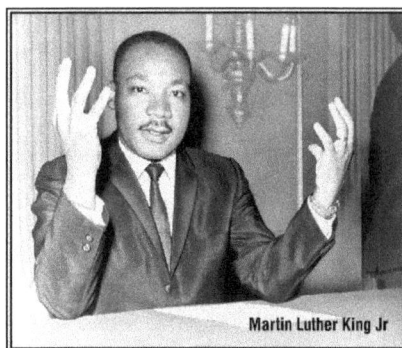

John Lennon   Martin Luther King Jr.

The implication of the understanding that Big Bang theory is incorrect and the Universe is indefinitely old is leading to fundamentally different prospects of the subject. In such an eternal Universe, the existence of more advanced civilized societies is more than certain. Our window for radio communication is about 100 years. In 50 years' time, we will have instant quantum communication, which probably the advanced societies are using. Realistic time to receive an answer with radio communications is counted by thousand of years. In 50y time, we will stop listening to the radio! I believe that the same scenario will be valid for every advancing civilization possessing radio communications.

The second reason is that primitive civilizations are very likely to destroy themselves. At the moment we are doing exactly this and rapidly approaching the point of no return with:

- Runaway global warming from melting methane layers.
- Unregulated genetic engineering.
- Quantum computer and artificial super-intelligence.
- Social limitations of incentives for human development and progress.
- Accidental or deliberate nuclear war, followed by biological warfare.

The third reason the universe is silent because the advanced civilizations are using quantum information to pass coded messages and information instantly anywhere in the universe. Our way of communication is open, cannot be coded properly; it is very slow and is not suitable for the vast distances of the universe. We are still very primitive, and with our actions, we are going towards inevitable extinction because we haven't got a social institution capable of stopping this! The real tragedy is that we do not possess any value that would convince others to give us help and guidance. We are just cruel, stupid, ignorant, and corrupt. We are laying each other on all levels and with everything we can. We are killing and robbing each other on a global scale! We are deceiving each other even with the vital questions for understanding the world - the origin of life and Big Bang Theory! The imposed on us limit of knowledge and communication will stay on until we prove that we are civilized and ethical society! It is obvious that we are not accepted and are not allowed to communicate with others!

I would like to explain to you what the implication is of the embedded limit of knowledge and the ethical limit for survival and existence of emerging civilization:

The laws of physics state that elementary particles are determined by their velocity and position. If we obtain and build super-computers capable of knowing all the elements of some parts of the universe, we could calculate what the development of this system will be; - in short, we will know the future of this system. If we have this capability, it will be easy to rearrange the matter of this part of the universe for our purpose. - Effectively we will have the capacity to arrange the future of selected parts of the universe. And to prevent this, the 'designer' of the universe has taken measures for this not to happen! - There is an embedded limit of knowledge, (conscious knowledge) where the increasing knowledge of one of the two basic properties of particles proportionately reduces our ability to know the other property. This dependency actually prevents us forever for being able to play and manipulate the properties of the matter and the universe. This is one of the limits for us embedded in the structure of matter, and is a very good one!

The other important limit is that quantum information, which is the actual mechanism of the implementation of the laws of physics and is also the information link between all particles of the universe, is coded with an unbreakable code. - This is the next mechanism which if we obtain, will give us the ability to make changes to the world or part of the universe! - And exactly these properties and functions of matter are effectively beyond our reach. – (Because the information is coded with an unbreakable code)!

The next obstacle (or limit) is that the quantum information is not part of matter, but is stored safely beyond our reach in the universal consciousness. The other fact (or physical limit) is when energy crystallizes in the form of matter, it always produces the same amount of matter and antimatter, which annihilates and goes back to energy. Effectively, this is an embedded mechanism as a law of physics, which is preventing us from creating material structures from energy! Those four carefully chosen measures cannot be coincidental! Nature does not create conditions or secret unbreakable codes. This is the biggest proof that the universe is designed by an intelligent mind with knowledge and vision for developing intelligent life, and with a full understanding that this emerging intelligent life can be corrupt, brutal, cruel and destructive as humans!

There is another crucial limit set by the creator of the world, - the final limit for our existence! And we are right on the edge of this limit. The logical features which pointing to the setting of ethical standards as a limit for civilizations are the fact that we are given freedom to obtain our last deadly inventions: - the atomic bombs, quantum computers, and genetic engineering. - These are the last three inventions of any corrupt civilization because the corrupted soonest or later will find the way to use them!

Those inventions can be controlled only by an ethical, unified, and well-organized society!

The quantum computer has the ability to use a small part of the universal intelligence. With only – about 500 entangled particles, the quantum computer will be billions of times more powerful than all existing computers on Earth.

Can you imagine the capability of the intelligence behind the scene that is currently using all the particles of the universe? We are stupid and arrogant enough to deny and not obey the rules set by such superior intelligence! Why? If we build a quantum computer, this computer inevitably will develop super-intelligence far more supreme than ours, and this super-intelligence would not necessarily see any value in our existence, because we are just brutal, corrupt and destructive. It is very logical in such a scenario that this intelligence would have to take care to liberate the nature of Earth from its destructive parasites – (humans)

Genetic engineering is another way of self-destructing if this capability is in wrong hands. Currently, these capabilities on planet Earth really are in the wrong hands!

If has a chance, the elite will use the advances in bio-engineering to make us quiet, obedient citizens, without any ambitions or desire for any advance, or sophistication! Currently, the elite are working to solve the problem - how the genetic mutations can be delivered by airborne viruses!

The logic behind allowing us to obtain this deadly capability is undeniable! - Each of those three capabilities in the hands of an irresponsible, cruel, and corrupt society will inevitably wipe them out. The catch-22 behind this scenario is that only an ethical, tolerant and united society can have intelligent and effective control over these dangerous inventions, and such control can be achieved only if society is able to live in absolute unity, peace, and harmony, with unbreakable principles of ethics and tolerance towards each other. - This is the bitter reality and the logic behind the ethical limit for developing intelligent societies and is the answer to the question: 'why the universe is silent.'

ECONOMICAL DESIGN OF MATTER, STABILITY OF ATOMS AND
THE LOGIC OF UNIVERSAL STRUCTURES

Admiration is the best description of the world in which we live in. The vast and endless universe provides us with never-ending possibilities for exploration and expansion of life. We not only have to admire the beauty of the world but have to understand its structure, design, and purpose. When we analyze how the world is constructed, it is not hard to see that logic is embedded in every detail, every element and function. There is logic not only in the physical structure of the universe, but it is also infiltrated deeply in living nature and eco-system.

Let us examine first the structure of matter. We know that raw nature is not economical, is not perfect in design and functions, that nature does not provide plants with the necessary drops of water, but creates droughts and floods - (too much or not enough); extreme cold and scorching heat. There is no logic when nature creates mass extinctions and acts in a chaotic and sometimes destructive way. In comparison with the chaotic pattern of behavior and functions of nature, the design of matter and universe in most aspects is absolutely perfect. The balance of all forces and properties is calculated with unprecedented precision. This incredible precision and balance of atomic forces give matter absolute stability. It is working like a Swiss watch!

Electrons orbit nuclei continuously for billions of years; and nothing can affect them, deplete their energy, stop them, or destroy the atoms or matter. - Moving charged particles must emit electromagnetic waves, which is the case with electrons. According to the laws of physics, electrons must collapse in the nucleus in a millionth of a second, but we do not observe such things!

The other unsolved puzzle with electrons in atoms is that they are charged particles, but they are not affected by any external electromagnetic forces! The magnetic fields of the universal bodies are interacting with each other and producing an electric current. And there again we are facing carefully designed special universal properties – the superconductivity of deep space allows those currents to transfer electrical charges and defused plasma on huge distances and act as the blood circulation of a living organism.

The combination of superconductivity and dynamic interaction of universal magnetic fields are providing a well-balanced system for cosmic matter and energy re-distribution and acts as a recycling mechanism of the universe. This energy distributing system providing balance to the universe and ensure its eternal dynamic existence. Such perfection in the design of matter is amazing, where only one substance – (energy) is used for the incredible variety of elements, minerals, chemicals, and substances. For the formation of all elements, mineral, chemical substances, plants, and biodiversity, are used only three particles: - electrons, protons, and neutrons. - Give three marbles to the smartest person on Earth, and he will not be able to make more than three combinations. But the unimaginable fact is that we have not millions of different substances, but billions of different products and building blocks in our world and all this unimaginable variety is made of only three basic particles! - Such ingenuity is beyond any imagination and claim for an accidental and spontaneous origin of the world with such ingenuity, calculated balance and logical purposes are absolutely not credible.

The ingenious way in which the heavy elements are created and distributed in the universe is a stroke of engineering perfection. According to the official model, heavy elements are produced inside the stars, and nothing can take them out. And if this is the case, we face the ingenuity of universal logic again! - The dying stars, after they have finished the job for which they were designed, very conveniently exploding as supernovae or start emitting defuse charged particles and bringing back the material in the convenient form of dust, heavy elements, plasma, and gases, which is the necessary building block for planets and biological life. The wonders of water are another marble of ingenuity! - Contrary to all other substances which contract when they freeze, water as ice expands. This tiny fact makes life possible because if the water had not been designed this way, in winter, ice would gradually accumulate from the bottom up in rivers and lakes and eventually the planet would freeze solid. When we observe the continuous logic and ingenuity in each element or structure, where thousands of forces, properties, the physical weight of the particles and forces are calculated and balanced with the precision up to 10^{123} - This cannot be regarded as an accident, because when such 'accidents' are piling over and over, we call such a chain of statistical coincidences - 'certainty'! This cannot be a freak accident, but it is a careful work of an engineering genius!

If we imagine that the atomic nucleus has the size of a pea, then the electron, which is unimaginably smaller than the nucleus, will orbit at a distance of 250 meters away from the nucleus. That makes an empty sphere of 500 meters in diameter, with only pea in the centre. And even this pea - (the nucleus) is not solid. - That is the real density of the matter, almost zero! Nothing! Empty! - Everything solid around us is actually made from cleverly balanced electromagnetic and atomic forces. Even the density of iron is 99.9999999999999% empty space. Don't these facts tell you something important? Are they not making sense to you? OK,...I will explain to you where the logic and sense in these amazing structures are:

Nature is not economical in any respect. We can say that nature is very generous and provides us with abundant resources: air, water, soil, trees, grass, and nearly everything with more than we need. In contrast to nature, when our engineers try to construct something, they always try to achieve the maximum structure, maximum strength, and reliability with minimum materials. - Every intelligent designer uses this principle! And we face the same logic in the design of the matter of our world. Just this design is so perfect that our intelligence will never come even close to such sophistication and perfection. Such perfection can be only being a product of absolute intelligence and unimaginable ingenuity. - This perfection revealing and is undeniable proof for its real intelligent origin!

When we are observing the structure of matter, we are facing undeniable logic and purpose in any detail of the microscopic world. It will be very unlikely this logic in the construction of the microscopic world not to continue also in the biggest universal structures. A bit to find the logic in the universal structures, we have to know its purpose! To find the purpose, we have to follow the hierarchy of importance, or in short - what is on the top of the ladder of the universal structures? – Obviously, it is consciousness! –The consciousness is on top, follow by the quantum information, the law of physics, then space, time and matter. And when we follow the structural ladder of importance will be easy to understand that the triumph (or purpose) of nature is the biological life gifted with consciousness intelligence! - And this is the logical purpose of the universe! – To create a variety of unique conscious, intelligent life gifted with high ethical and moral values!

Only Milky Way has not less than 100 billion Earth-like planets capable of harbor life! Do you believe that they are not thriving with life when we know that the earth's life has started immediately after it surfaces cool down enough?

Let now see how the bigger structures of the universe are created and are there also embedded sense and logic to serve the purpose of emerging intelligent life.

A well-known fact is that in every point of our sky is a star or galaxy. If the property of the universe hasn't been tuned precisely how they are and the intensity of light is not diminishing by the distance, we will live in the incredibly luminous world, where the blinding light will come from everywhere. This light will burn us! We won't be able to see anything, and simply, life won't be possible!

The officially accepted understanding of the world is that gravity is the dominant force of the universe. Again, the strength of gravity is balanced with such incredible precision that matter forms stars, planets, and life without crushing them. If the strength of gravity were just a little bit more or a little bit less, nothing would happen, and we will not be here. This strength of gravity allows the stars to be far away from each other, but still to be connected and to be in the visible range. What does this celestial arrangement mean to us? By any logic, this arrangement makes perfect sense for the rising and establishment of intelligent life in the universe. The big distances isolate emerging intelligent life, but this isolation allows freedom of self-determination, independent development without foreign intervention and the unique quality of every developing intelligent civilization. On the other hand, these big distances effectively protect not fully developed and non-ethical civilizations from exchanging stupid and corrupt ideas, preventing invasions, destruction, and war between them.

Is such an arrangement not serving the purpose and logic of creating perfect conditions for developing intelligent life?

However, those distances are not an obstacle for the ethical and advanced universal community that uses the instant speed of quantum information for communications. The ethical limit in the transformation from undeveloped civilization to develop is the critical element of the universal design. The condition of ethics is embedded in the reality and properties of matter, and they are an unconditional barrier against further development and the spreading of corrupt and unethical societies. This ethical limit makes sure that any unethical civilization will perish before being able to obtain the capability to travel and communicate over big distances with the capability to colonize, exploit, and harm other civilizations.

When we put together all those logical elements of universal design and incorporate there the findings of quantum mechanics, it is obvious, that the entire universe is designed with one simple universal formula! And this formula is providing a system of balance, certainty, continuation and eternity.

It is time for us to realize that the universe is designed with great vision and ingenuity far more superior to our best imagination!

The statement of astronomers that the universe is homogeneous on a grand scale, providing us with the knowledge that the universe acts as a closed physical system.

And in reality, the universe is an enormous closed physical system, where the stability of space providing the basic foundation for the other components of the universe to exchange values and the system to work. The energy inserted in the universe has nowhere else to go! This arrangement is providing the eternity of universal structures! The precision balance of forces and energy interchange between the two dominant forces - gravity and electromagnetism are providing the dynamic mechanism for energy and matter exchange, and this is the actual continuous recycling and re-juvenile system of the universal structures! Is time to realize, that this is the common universal formula (or system) embedded also in every natural structure and in every single organism. Our bodies basically have the same design! - Our consciousness is producing the information, which allows us to thing and regulate our body to function! In our cells is imbedded the biological information of DNA, which works on the same principles as the quantum information of matter! The universal structure gives us the freedom to grow, to exploit, and to have unique cultures and absolute securities with the assurance that nobody will harm or colonize us, but these privileges we are given coming with special conditions attached! Those measures are applied to us too! There are imbedded measures ensuring that we also will not harm others! –We have to understand that the Universe is not a free lunch! We are part of this system and have to obey its rules! The choice is ours - obey them, or perish!

The simple formula of the system of the universe is the best example for us to follow because this system is proven by the existence and the harmony of universe and nature!- It works!

Won't be difficult to adapt this formula and system, where the fundamental principles are unchangeable, and where is embedded effective safeguard against harmful acts in the form of limits of knowledge and necessity of ethical standard!

This simple formula will provide us stable democracy, harmony, prosperity, and a future!

How we can adopt the principles of the universal structure as our social system, we will consider in the latest chapters.

The most important knowledge for humanity always has been the knowledge and understanding of our world. These fascinating questions always have been the most elusive and hard to answer, because to answer, we have to poses enough knowledge and to have the courage to make sense of the available facts.

In the previous chapters, we have considered all elements of the universal structures, and now we have to put them together to find out how actually the universe works.

Where the problem is coming from

The 'Standard Model' of astrophysics reluctantly has departed from the practice of using correct scientific facts and methods in pursuit of providing a scientific explanation of our world. The leading theories are full of bizarre assumptions, which violate the laws of Physics, observational data, facts, and healthy logic.

The scientific instruments are getting more and more sophisticated and are providing us with a rich amount of data and observational evidence for the structure of our world. But no matter how accurate is the collected data, if this data is not considered correctly with care and logic, the avalanche of new information will create more confusion rather than explain anything.

Ultimately this is exactly what happens! The efforts of the scientific elite to keep the old theories in place are leading to ignorance of a range of crucial scientific facts and observations for building a correct picture of our world! Many scientists are stating that we have a crisis in science, but not one of them dares to identify what it is. So... why are they talking of a crisis and never reveal what actually the problem is, where it is coming from and what we should do to correct it?

I will start explaining what the crisis is with a few well-known facts:

100 years ago, when the fundamental theories for the structure of the Universe have been established, the scientists had no idea, that the visible part of the Universe is less than 1% of its real matter content.

With the development of radio-astronomy, we have discovered, that the diffused matter, dust and plasma in the interstellar medium represent at least 99% of the matter content of the Universe. - (This assumption does not include the hypothetical Dark Matter and Dark Force). – See the image below.

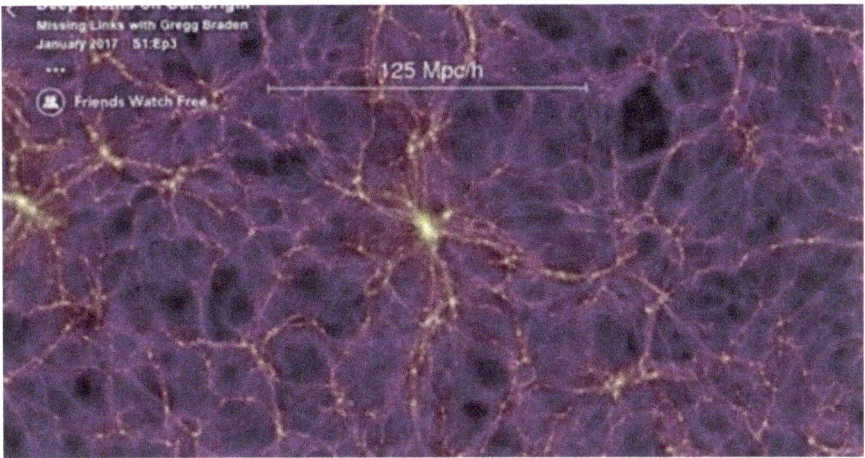

This is how the "Empty" space looks through the eye of Chandra space telescope. Why do we continue ignoring the existence of the "invisible" to eye interstellar filaments?

From the beginning of the 20th century until now, the foundation of astrophysics is still based only on this 1% of the visible matter content and is not taking into account the superior electromagnetic force, which is ruling the rest of the 99% universal matter and influence to great extent the energy exchange and movement of the visible part of the Universe. The electromagnetic force is many billion times stronger than gravity, and if we would like to have a realistic understanding of the Universe, we should include all those components and forces in our model.

Unfortunately, the current situation in physics is similar to a scenario, where we are given a microscope, to find the sense and beauty of a fully covered Rembrandt painting. Currently, we are studying the Universe with the biggest microscope on Earth – LHC. We are smashing particles with the hope to find and explain the picture and dynamics of the colossal universal structures without taking into account the 99% of matter and its dynamics! As a result of this approach, we are bombarded with bizarre theories for non-existing phenomena's as mass-less particles, Dark Matter, Dark force, Higgs boson, expanding Universe, Neutron stars, Black holes, Dark light and mysteriously "self-controlled" nuclear fusion of the Sun. We are spending enormous manpower and resources needlessly to keep in place old and impossible theories, which have nothing to do with the reality of the World! I just would like to mention, that all those puzzles are created by ignoring facts, observations and the law of Physics. In the following article – 'THE UNIVRESE AND THE THEORY OF EVERYTHING' such ways of selective considerations of the scientific facts will be avoided. The careful consideration and logical analysis of all available facts is leading to completely different results, different conclusions and a completely different picture of our amazing Universe.

THEORY OF THE UNIVERSE AND THE THEORY OF EVERYTHING IN PHYSICS (TOE)

Nothing has fascinated more the imagination of humanity as the burning question for the origin and secrets of our Universe. The Universe always has been and still is the biggest unexplained mystery in our life! The ancient philosophers and religions are giving us a very different picture and explanation of the Universe. The first models start with naïve pictures as turtles or whales holding the floating land. These funny models have been followed by the Geo-Centric model and replaced with the Helios-Centric and with the current model of The Big Bang Universe. Each time our knowledge takes a step forward, The Universe keeps surprising us, and each time increases its size in parallel with our knowledge. The current officially accepted model of The Universe has been created over a century ago and has a radius of 13.7by. The increased sophistication of our instruments lets us accumulate unimaginable amounts of data, which disagree with this scenario, but the model of The Big Bang theory still resists going away.

Is this really the final word of our science? Is The Universe really cannot surprise us anymore, or will there be more surprises to come? Is The Universe not ready again to reveal its new size of colossal scale and grand structures beyond our imaginations? Let see what the new data and observations will reveal to us.

The experimental and observational data we collect is enormous and is full of facts that cannot be accommodated and explained by our current Big Bang model anymore. Even Edwin Hubble, who is instrumental for the creation of this model, wrote a letter to the US Astronomical Society denouncing his support for the theory which made him famous! - He was really a brave and honest man!

I don't like to be drawn into baseless arguments. The explanations of the fact. I am presenting and the facts for the impossibility of The Big Bang theory can be found in my book 'Myths Lies Illusions and The Way Out.' I just would like to mention that the average intelligent person can judge the correctness and implications of the following facts:

- ❖ On the base of one single false assumption that light red-shift indicating stars recession we are forced to admit an impossible scenario. - Even Edwin Hubble, the man who becomes famous for this discovery later has denounced the credibility and correctness of this assumption.

- ❖ If the Universe comes out in form of explosion, the result will be just a cloud of pure energy or a bunch of incompatible particles. - No one explosion is able to create sophisticated logically build structures! - This is an undeniable fact, because - we humans have managed to explode billions of all kinds of bombs, and we never observe that they are able to create anything! - (Such scenario as explanation is not credible)

- ❖ If there has been a Big Bang, the galaxies and stars toward the Big Bang location would have to get closer and closer and their density must turn into a continuous bright curtain just in front of this event- (Big Bang). Contrary to this scenario, our sophisticated telescopes reveal a continuous homogeneity of the celestial structures in all directions and all distances! Everybody can check this fact because the Hubble 'Deep Field' pictures are available on the internet!

- ❖ The second undeniable observational fact is that the observable Microwave Background Radiation (CMB) is coming from the opposite direction of the proposed scenario of The Big Bang! - This radiation is a sphere with a radius of 13.7 billion light-years and is on the outer boundary of the observable Universe. This radiation has been emitted 13.7 billion years ago from this enormous sphere on the edge of our observable universe in time when the universe is "supposed to be" smaller than an atom in the centre of this sphere, and the CMB is supposed to be emitted in the centre and to travel out toward the edge of the observable Universe, not in the opposite direction! This is undeniable observational evidence, which proves beyond any doubt that 13.7 billion years ago The Universe has been the same size as it is now! In addition to this, the combined time for the expansion and then the traveling back of CMB is at least double - (27.4by) than the proposed age of the Universe of 13.7by. –

- ❖ (See the graphs - pages 48 -54)

It is time to leave the nonsensical arguments behind and start using the available accumulated data and evidence to draw one more realistic and correct picture of our Universe!

I would like to explain that according to the scientific standards, for a theory to be accepted as scientific, it has to be based on realistic data and evidence and not be in direct violation with the laws of physics. Usually, when some theoretical predictions fit the available data and facts, and not leave any unexplained anomalies, which have to be explained with mystical assumption such as 'Dark Matter' or 'Dark Force' then this theory has all the qualities to be correct. This is the scientific standard that I am strictly applying to my explanation in the aim to solve the puzzle and the mystery of the Universe! I will reveal the facts related to the structure of the Universe and refuse to be drawn into further baseless arguments or speculations.

If the readers are not sure and need further proof of those facts, they are welcome to purchase the book mentioned above, where the correctness of all the details are explained in easy to understand language.

THE KNOWN FUNDAMENTAL COMPONENTS OF THE UNIVERSE ARE:

The Universe is constructed of material and non-material components:
Space, Time, and Matter is the physical (material) components of the Universe.
Consciousness, Universal Information and the Law of Physics are non-material components of the Universe.
The scientific community is considering only the material aspect of the Universe.
The religious community is considering only the non-material aspect of the Universe.
Consideration of only half of the building blocks of the Universe has lead to Materialism and Spiritualism.
Both philosophies are a primitive and incomplete way to deal and explain the properties of the Universe!

- CONSCIOUSNESS – The Universal Consciousness is not a material substance and is not affected by the physical processes of the Universe. Consciousness is "symbioses" of intelligent information and unknown to us a form of energy. The origin and structure of consciousness currently are unknown, but its presence is everywhere, it is infiltrated into every particle, and every living cell! You cannot create, destroy, duplicate, or manipulate Consciousness! Consciousness is the first fundamental phenomenon of the Universe and is responsible for its existence, and the existences of all other components of the Universe. Consciousness is the origin of the Laws of Physics. It is the memory storage and origin of the Quantum Information, which is the actual tool or is the physical mechanism of the Laws of Physics to be implemented into every part and every particle of the Universe. In association with our computers, the Consciousness is correspondent to the hard drive, and the Law of Physics is the window software. Quantum information is a product of those two components we just mentioned. Consciousness is not a product of our mind! The fact that the plants and all living organisms possess consciousness, and most of them have no brain, or nervous system is the actual proof that our consciousness also is not a product of our brain! Our mind is just our processor, which is using consciousness as every other organism and every part of the Universe! The capacity of the "processor" determines the quantity of consciousness it can possess and use.

- THE LAW OF PHYSICS – The safeguard mechanisms of the Universe (explained in my book) is not allowing us to know and control the fundamental properties of the Universe. The law of physics is one of them. The limit of knowledge is not allowing us to know where the Law of Physics is situated and how it is controlling the quantum information and the physical processes of the Universe. The laws of physics are not material substances and cannot be affected by any physical processes such as heat, pressure, gravity... The Laws of Physics are a sophisticated informational order which can be a product only of an intelligent creative mind, or phenomenon. The laws of Physics precede the existence of the material components of the Universe because they cannot exist without physical order. This logical order reveals the consecutive design pattern of the Universal structures. The Law of Physics is acting as the 'software of the Universe'- It is the intellectual product of the Universal Consciousness; the Law of Physics is a designed plan and functional order of The Universe!

- QUANTUM INFORMATION – for simplicity I am using this term to describe the Universal Informational link between all parts of matter and structures of the Universe. Quantum information is not a physical phenomenon! For this reason, the physical processes of the Universe do not affect the information! The law of Physics states that you cannot create, destroy, copy or duplicate quantum information! Quantum information is the informational link between each particle and each physical structure of The Universe! The Quantum

Information travels instantly at any distance and is not affected by Time, because Time is energy-related substance and belongs to the physical part of the Universe. The information is the invisible tool or is the "hand" of the Law of Physics and is the mechanism to keep the order of The Universe! The quantum information is stored in the Universal Consciousness! The information does not belong and is not stored in the matter – (How is the official scientific view). This fact explains why and how the particles are able to "remember" their original property and spin when they change their physical forms!

- TIME – Is a single dimension – it is a vector, dynamic progressive direction without volume, without beginning or end.

- SPACE – Our space is a **six-dimensional** physical phenomenon, (or component) of The Universe! - (The explanation of space is on page 12). As everything possesses physical property in The Universe, Space is also made by the common building substance – energy! The three physical components of The Universe are **Space, Time, and Matter**. These fundamental physical components share a common building element, which is **energy**. Energy and information are the unifying elements of the material structure of The Universe!

What is a single dimension?

To be able to understand the structure of the World we need to have a clear understanding of the most fundamental elements, which are the building blocks of the Universe. - Space, Time and Matter. Our science still is struggling with the formulation of these basic elements.

I will start with consideration of our understanding of spatial dimensions and particularly with our understanding of what a single space dimension is. There are conflicting statements about the nature and property of a single spatial dimension as "Higher dimensions", "Inductive dimensions", "Tangled dimensions", "Minkowski dimension", Hausdorff dimension" and ... endless range of scientific speculations. We have a very good example of the nature of a single dimensions – which is the Time!

The Time is described as an arrow, (or vector), but nobody is taking real notice of this. If we apply this knowledge to the understanding of what exactly is a single spatial Dimension it won't be difficult to realize that Single Spatial Dimension is a Vector, it is <u>dynamic Direction</u>! It is a progressive directionally oriented phenomenon. The assumption of the creators of 'String Theory' that their "Strings" are one dimensional is a clear indication that present science doesn't understand what is a single dimension! - Their "Strings" suppose to be material objects. - Any material object will have volume and will be three dimensional no matter how small it is! On top of this, the "strings" have a specific length and two ends. - Any object with two ends can be measured in two directions, but a single dimension cannot be measured in reverse! - A single dimension is a one-directional vector. (See the diagram page 10)

Any physical object or material substance also cannot be two dimensional, because the matter inevitably has volume! The two-dimensional configuration has only surface without volume and also cannot accommodate physical objects!

The next fatal flaw of modern science in understanding space is the lack of understanding that a vector can be described as a range of progressive line of consecutive reference points. The reference points in space can be described as - (Here and Now) (The detailed explanation of "Reference point" is a few pages below).

The next fatal flaw of modern science is that they always is putting the reference point out of considered surface or space volume.   (See the diagrams below)

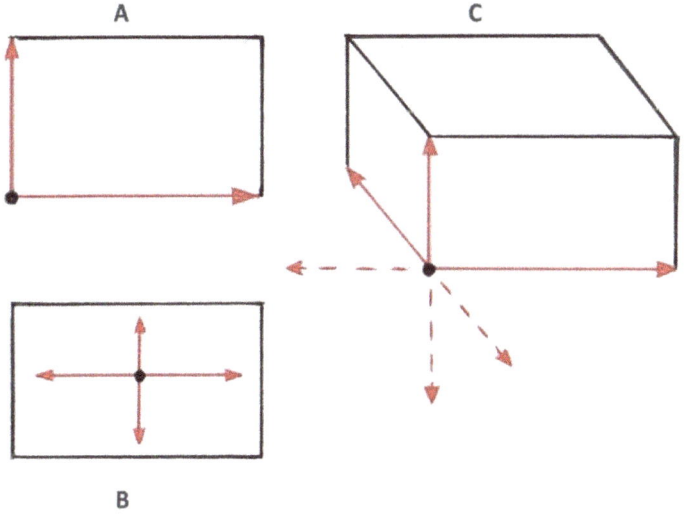

A

C

B

On the diagrams "A" and "C" is visible, that the reference point is outside – it is on the other boundary of the considered objects - (surface and volume). In the Universe is no such place, where the reference points can be situated outside! - Everything is inside in the volume of space!

To describe a flat surface the reference point must be into the considered surface! And to describe a surface is obvious that we need minimum four

146

<u>vectors</u>!

From ancient times, we know this! We have drawn millions of maps and our Sea navigation is based on the principle of Four Directions! - East, West, North, and South! (See the diagram "B").

Unfortunately, the scientists didn't get notice of this fact and succeed to make a real scientific mess with the assumption for the existence of only Three Spatial Dimensions. - See the diagram "C" where is visible, that three dimensions can define only 1/8 of the total volume of Space because Spatial Dimensions have only one direction and cannot be measured in reverse!

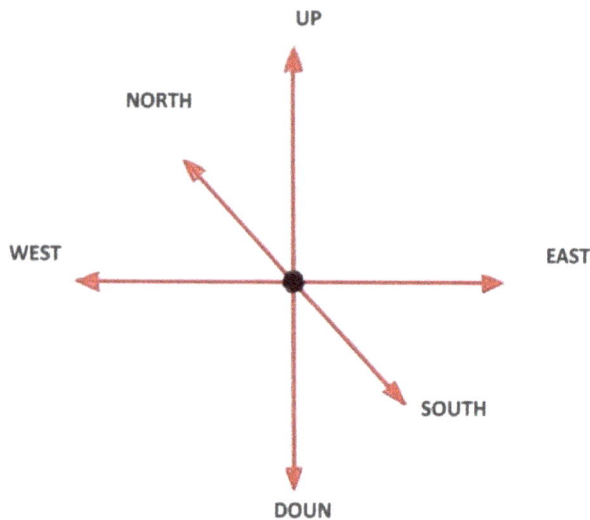

UP

NORTH

WEST   EAST

SOUTH

DOUN

From every point in Space, we can travel in Six Directions! Three-dimensional configuration can describe objects with finite space volume! Three dimensions cannot describe Space! We have to understand the difference that <u>Every physical object is three dimensional, but space is six-dimensional</u>!

From every reference point in Space is beginning Six spatial directions - East, West, North, South, Up, and Down!

TIME - what it is and how it works

Time is a very difficult subject even to be described. The time delusion is giving us the understanding that time is energy-related and belongs to the physical part of the Universe. So far, nobody has come even close to find the physical constituencies and physical property of Time. There is a logical pattern of limits imbedded in the property of matter and the laws of physics, which prevents us from obtaining exactly the knowledge, which will give us the ability to mess with the order of the Universe. It is more than obvious that those limits are there for a good reason and they are not accidentally inserted! The knowledge of time's property is the next area, where the imbedded chain of knowledge restrictions definitely should be implemented! And exactly this is the reason - the physical property of Time is to remain a hidden and well-guarded secret. The limit of knowledge is preventing us from knowing the exact physical property of time but is not preventing us from knowing how time works, and how the mechanism of the irreversibility of the physical processes in the Universe works.

Everybody is talking about the arrow of time, but nobody has yet realized what exactly it means and been able to resolve the Mystery of Time. Einstein included time in his theory but wasn't able to explain why and how the time makes the processes irreversible. Scientists are looking into the microscopic property of matter to find it. They are calculating and manipulating numbers relating to space, time, dimensions and creating bizarre theories, but they are not giving us any reasonable and logical explanations of this puzzle. The invented mathematical formulas with adjusted numbers and values do not make sense and is a road to nowhere! These formulations are leading to bizarre theories, wormholes, parallel universes, time traveling and much greater fantasies that are good for Hollywood but are without any concrete evidence and do not provide any credible solution and explanation of Time.

I believe that I have the answer and will be able to unlock the secret of time mechanism and give you a good and easy to understand a credible explanation, by using logical consideration of the available facts, and the known relations between the fundamental properties of the Universe.

As usual, most of the genius inventions are simple. We have observed the economical and simple pattern of design in every part of the Universe and its structures. This criterion of simplicity is applied to the Time mechanism too.

To understand how the universal physical system works and how Time is producing the irreversibility of physical processes of the Universe, we have to start with consideration of the correct physical property of the fundamental elements of Universe:

The matter of The Universe is situated not in three, but in <u>six-dimensional space</u> where everything can move freely in all six directions. The volume of our space is proportional to the amount of matter. We have to understand the fundamental difference - that <u>every physical object is three dimensional, but space is six-dimensional</u>! The assumption of string theorists that their strings are one dimensional clearly shows their complete misunderstanding of what actually dimensions are! They are measuring and adjusting the string length to the Planck Constant! - Any physical object with two ends cannot be one dimensional! A single dimension is a <u>one-directional vector</u>. Any physical object or material substance also cannot be two dimensional, because the matter inevitably has volume! The two-dimensional configuration has only surface without volume and also cannot accommodate physical objects! Currently, the only known one-dimensional phenomenon is –Time! And this is giving us an understanding of what a single dimension is.

This is the unique property of Time! - That time is one dimensional directional oriented progressive energy field – (this is a well-known fact), but really, this "fact" is not well understood, is not formulated correctly and is not taken correct logical conclusions. Time is energy-related and is part of the physical aspect of The Universe! Time is dynamic and varies, but Space is steady and constant! A well-known fact is that space is not affecting time, and time is not affecting space eider. These facts are telling us clearly, that Space and Time are not physically incorporated! - And exactly this is the great miss-understood physical concept of Space and Time relations. - The fact is that Space and Time are part of the same system, but they are not physically incorporated!

To understand Time, we have to understand first what one dimension is! - To have a single dimension means that the single dimension is a vector and has only one direction! Time is a single dimension and cannot be measured in reverse! <u>Time is a progressive, dynamic direction</u>! This is the answer to the puzzle, and this is the real meaning of what a single dimension is: - Single dimension is direction only! A Single dimension is a vector! It is an arrow and is direction without volume! A single dimension has no point of beginning or point of end! It is a direction without volume! <u>Time is a progressive, dynamic directional physical phenomenon, which applies to every part of the Universe</u>. **Understanding the nature of a single dimension is the key to understand the entire Universe!**

To be able to explain how time works; I am providing the drawing below, where it is visually explained the difference between three, two, and one dimension. Also, in the drawing, we can follow the actual process of irreversibility, which the single time dimension (or the vector of time) is providing to The Universe.

3. Dimensions (Our Space)
2. Dimensions (Flat Surface)
1. Dimensions (Direction without volume (time))

TIME DIMENSION

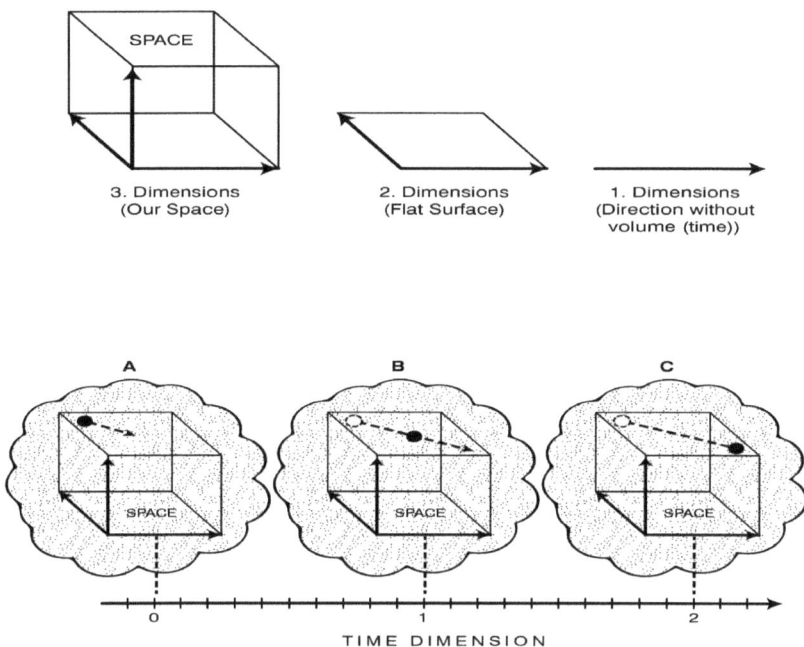

In the diagram, it is obvious that the black object in frame 'C' cannot come back to its original position in the opposite corner, because its original position is in the opposite corner of frame 'A' but the "universe" has already moved from point '0' to point '2' (or from frame 'A" to frame 'C'). – (like a ratchet)! So... there is no way back!

That means that every new second, everything is situated in a different place of the one-directional <u>Time Dimension</u>. The continuously changing (space position) on time dimension and the irreversibility of time direction effectively locks up the physical processes (like ratchet) and makes them fixed and irreversible, because nothing can go back to its original or previous position, where they have been in the past! Effectively, time is moving everything continuously in its one-dimensional space direction! This is how time works and is an explanation of the arrow of time and is the answer to the puzzle of how time provides the irreversibility of all physical processes and everything in The Universe! It is a simple explanation of basic fundamental physical interactions, which works simultaneously for everything which exists in The Universe - from subatomic particles to the biggest structures of the Universe!

Objects space position - reference point and its dependency of Time:

It is obvious that Einstein's concept of Time relativity is incorrect because when we apply its concept we are running in absurd situations as the "Twins paradox" and the "Photon's relativistic time freezing."

To ignore or adjust the absurd results to fit the incorrect concepts is not a good way to do science.

Every object has it's unique position in Space and Time. We will consider what that means.

Time is included in defying the position of all objects in the Universe. I will start the explanation with two points in space - our position which we calling **"Here and Now"** and another point we calling **"There and Now"** ("Here" or "There" is the object's position in Space, but Now is its position in Time).

Time has only two specific properties: "Now" and the "Future".

Every point of space can be a reference point and be "Now" from which the time and space is propagated out toward the future in all directions. That means that every point in space has a different time value!

(See the graph below)

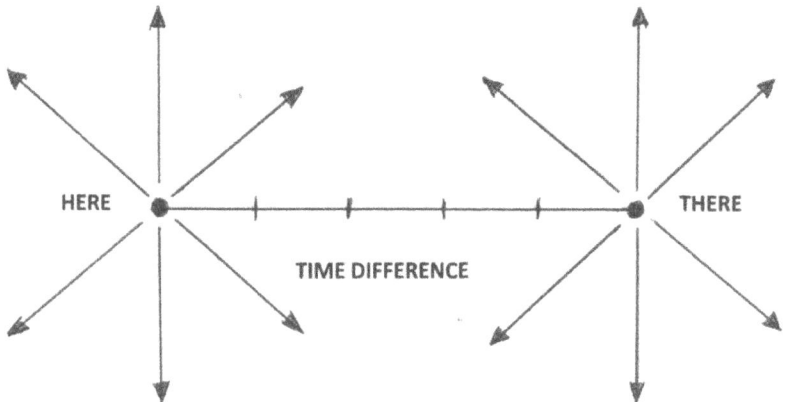

(Every physical object has its specific point in the Time Dimension)

For easy understanding I will give you an example - if you are standing on the North Pole there all directions are South. If you are standing on the South Pole, all directions will be only North. Same phenomenon we are observing with the two positions on the diagram "Here" and "There". That means that every point of space could be a reference point, where the Time is propagated outwards and all directions are toward the future. Every point of space could be "Now" but "Here" and "There" are located in different places of the Time arrow. From this example, we are reaching the understanding that every point of space has its unique location because is situated in a different place in Time and Space Dimensions. Or in short, the Time is not change its speed (how the relativity proposing) the <u>Time speed is constant</u>, just the Time and Space dimensions are crossing each other in different positions of their vectors. For slow-moving material objects, this time difference is negligible, but for the fast-moving quantum objects as particles or waves which is carrying with them the information for their local time value, the time difference becomes obvious. <u>For the "Receiver" the incoming waves will carry more advanced time</u>. This is the currently observed in GPS so-called "Relativistic" Time advance. This is what we observing in Mercury orbit precession, <u>(not recession) which the Theory of Relativity predicts</u>. The difference in time values of points in space is because space is steady, but Time is an active progressive dimension. Time is a progressive directionally oriented energy field. Currently, we not have any observations that Space, Matter or any energy field is able to interact and change the Time value. We need to find the Time specific energy density value to be able to construct a precise picture of the relations and the results of the interactions of Time and Space. Time and space are separate physical entities! Space is steady, but Time is active and progressive. - And exactly these separate physical properties are providing the mechanism for the irreversibility and position separation of every object in Space!

In the argument: is the Time absolute, or relativistic, the absolute Time is more credible because if the relativistic time is ruling our world, the top of the mountains and skyscrapers will slowly drift back and lean toward the west, because of the alleged Time difference provided by the earth's rotation. The provided data is not reliable and there still is room for discussions and arguments with this subject.

We are measuring Time with our clocks, but we do not measure the length of Time! We are measuring its' speed only! The speed of Time is related to the intensity and momentum of its energy field!

I am afraid that I will disappoint many people, but we are coming to the point of understanding that the fantasies for Worm Holes and travelling to the past are baseless and impossible! It is prohibited by the laws of physics because Time is irreversible! **Time is one-directional by its physical nature of being a One-dimensional vector without beginning or end!**

How time applies to position, velocity, and uncertainty as defying factor:

From the "Theory of Everything" we know, that Time mechanism is applied simultaneously to all parts of the Universe and is providing the irreversibility of all physical processes. (See the graph page 146)

To be able to understand the principle of uncertainty, we have to understand the crucial role of Space and Time in defying the object position and velocity. We know and are easy to understand that objects can have a different position in Space, but currently is unknown, that **the different positions in space also is different positions in Time dimension**. Above we have defined velocity as the Time difference between two points in space. The object position we have defined as a defining point in space and time. - It is obvious, that to measure the velocity of an object we need to know two values of Time and their difference, but to define the exact position we have to know the precise value of time in a precise point of space! (See graph pg 13) These two demands are conflicting, because we have only "One" Time dimension, and we cannot apply simultaneously three different time values on the same object, because if we use precise Time, we can define the precise position of the object, but we are losing sight of its velocity. And the opposite - if we apply to the object the time difference between the two points in order to define its velocity, we cannot apply the precise time to determine simultaneously its position! The fact is that we cannot apply three different Time values on the same object **at the same time**! - This is the phenomenon, which is behind the "Puzzle" of the Uncertainty principle. The fact, that we are dealing with Time to define the velocity or the position of the objects is stripping us of any possibility to know simultaneously the velocity and the position of the elementary particles – There are no "puzzles" if you understand the physical processes!

UNDERSTANDING SPACE

The final configuration of space properties currently is beyond our reach. Fortunately, the interactions of the fundamental forces and elements of the Universe is giving us a good base for understanding the fundamental structure of our Space, that Space is medium, containing all known to us forms of energy fields. It is similar to the white light, which contains all the colors in it. To understand space, we also have to consider the other two physical

components of The Universe (Time and Matter) and their fundamental relations with Space. We have to start with the understanding that space is an energy-related physical phenomenon. For a specific amount of matter (energy=matter), there is allocated a specific (proportional amount), or volume of Space - (because the Universe is homogeneous on large scale). Space and matter are in strict proportional dependency! That means that Space is a constant and unchangeable fundamental component of the Universe because we know that the amount of matter is constant! The constant volume of Space is providing the stability of the physical system of the Universe. This structure is providing the condition of a closed physical system of The Universe – regardless of is the Universe is finite, or endless - the energy and all physical processes have nowhere else to go! - This is the key factor behind the finely balanced matter-energy exchange of The Universe. The closed physical system of The Universe is providing conditions for the endless cycle of energy-matter exchange and all the universal structures to be continuously reborn and re-juveniles. This provides the eternal dynamic existence of our Universe. Space is a physical medium with specific physical properties, and the ability to propagate waves. Space is a perfect physical medium, with the conductivity range from zero to superconductivity, dependent on temperature, matter density, and intensity of the fields involved. Space is acting as a real physical medium containing an enormous energy field, which provides the different energy levels for the attractive forces mechanism; giving finite speed to traveling particles, and allows propagation of the electromagnetic waves - (Light). The observational evidence for this is all around us. Also, the revealing fact is that the passing waves and particles are losing energy – (The Universe is not blindly luminous). We know that in every point of the sky is a star, and if the light is not absorbed, the sky will be blindly luminous, like we are surrounded by a sphere of billions of blinding suns! The science has to find the exact rate at which gravity and electromagnetic fields are depleting the light energy and what role Space is taking in this process. Good observational evidence for the light depletion is the Cosmic Microwave Background Radiation. This radiation is the shine of the distant galaxies of The Universe and is coming from the distance beyond our
observational range of (the visible Universe). The energy of this light has been depleted below the visible range of the light spectrum - to the range of microwaves! (See the diagram on page 26)
The assumption that our space is three dimensional is not credible! In the

previous chapter, we have considered the physical nature of the single dimension, which is progressive, dynamic direction only, without a beginning or end and without volume! The assumption of the leading scientists that the description of the World has to be in the form of a purely mathematical formula is not correct. For example - the mathematicians are describing surface as two-dimensional phenomena, where two vectors can define any point of the surface. The difference between this mathematical assumption and the real world is that the mathematicians putting the reference point of their vectors always in the corner (which is out of the measured surface!). But in the real world (as a map) we and our reference point are in the middle of the map. - To define a surface, we need not two vectors, but four to be able to define each point of this surface or map - <u>or we need four vectors (or directions) – east, west, north, and south!</u> – This is the difference between baseless mathematical assumptions and the reality of the physical World. The same fundamental mistake is made with the assumption that we can define the volume of space with three vectors only, because the mathematicians again putting the reference point of their vectors in the corner (of the measured box), which is out of the defined space volume! I am sorry, but there is no corner in the Universe! - The Universe is all around us! Every point of the Universe is a reference point for defining the space around it! There are no corners in space <u>and there are not three, but are six space directions!</u>

The crucial mistake of the current three-dimensional configuration of space is the wrong assumption, that from the reference point "they" can go in the opposite direction of the three vectors (or space dimensions) to define the rest of the space, but simply this is impossible because each space dimension have only one direction! We have to realize, that each direction starts from the reference point and is a separate dimension! Each reference point in space is "Here and Now". We are perfectly designed to feel and sense the real World. There is no better and more credible description of the World, than the picture, which our senses are providing us. We know that the Cosmos is uniform in all directions. We know that there is only one Time Dimension. We also know that the World has six directions; for example - 'West' is just a direction, it has no volume, no beginning, no end, but is real, is existing, and we know that! We also know, that there are another 5 space directions! East, North, South, Up, and Down. We wouldn't know for the existence of the six world directions if they really not exist! We have to realize also the fact that each world's direction exists because it is representing a separate dimension.

They are six directions, we know them, and prove for it is that we can travel in each direction from each point of space. <u>To define the space around us, we need not three, but six directional vectors</u>! Three space directions (Vectors) are not enough to define the whole volume of space. In the diagram below, it is visible that three dimensions are able to define only 1/8 of the total space volume! Three dimensions will produce a disproportional configuration (distortion) of space and an uneven continuous expansion because three vectors are defining only 1/8th of the total volume of space, and their dynamic momentum remains active and cannot be restricted of continuation. progress. - (See the diagram on the next page)

Now we have to consider the subject of space dimensions and how they are incorporated together:

Correct understanding of dimension is that <u>the dimensions are progressive directional oriented energy fields; they are physical components, which have dynamic directional momentum and define energy intensity.</u> The correct understanding of the physical property of space requires a total reconsideration of our understanding of Space, Time and Universe. The current official assumption for our space suffers a lack of understanding of what really the dimensions are! - The dimensions are directions, which cannot be measured in reverse! This understanding explains the flaw of three-dimensional assumptions for our space. Our space cannot be three dimensional only, because three dimensions are not representing the total volume of space, and will produce a fatal space distortion! We have to understand the fundamental difference, that the physical objects are three dimensional and they are described by three vectors with defined length, but, <u>space cannot be formed and described by define length vectors, because space dimensions have **progressive directional energy momentum** and because there are six separate space directions</u> – East, West, North, South, Up, and Down, <u>and for each space direction to exist they must be represented by a separate dimension!</u> It is vital to understand the difference: that three-dimensional vectors are defining only 1/8th of the total volume of space! And in the three-dimensional space scenario, the progressive dynamic directions of each dimension will remain active! This unrestricted dynamic momentum will distort (expand) the 1/8th space defined by three active separate dimensions, which will continue expanding forever, reducing space energy density and diminishing the matter at a very fast catastrophic rate! **Three active dimensions <u>are not enough</u> – they cannot produce the steady, uniform and balanced space, which we are observing and enjoying!**

In order to have steady, and uniform space, **the active directional momentum of each dimension must be cancelled!-** And they are cancelled with the opposite directional dimension! - (East is cancelling with West, North with South, and Up with Down). The dynamic momentums of the dimensions are canceled, but their energy fields remain in a steady-state and preserving its space orientations! - (See the drawings below)

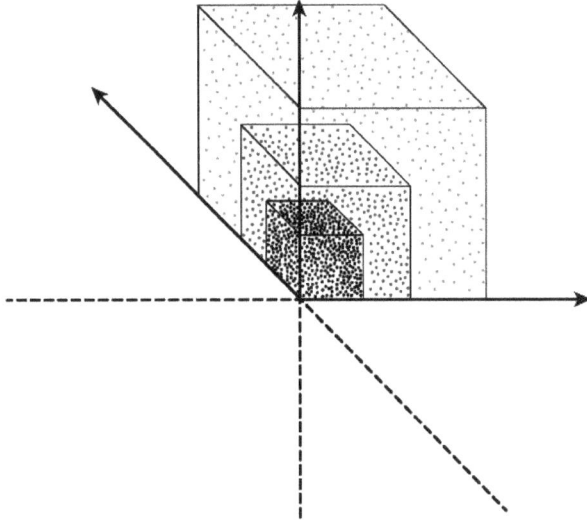

3D vectors producing only 1/8 of Space volume and will create continuous catastrophic distortional space expansion

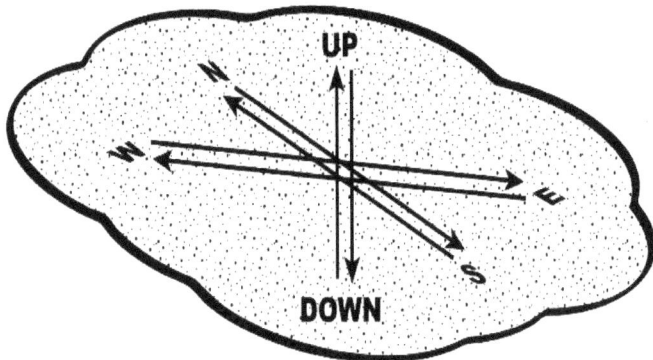

The 3 pairs of space dimensions are cancelling their dynamic momentum and producing uniform steady space and releasing energy = (matter) This is exactly the phenomena, which is producing the steady, define volume of space full with enormous energy, - it is the uniform space, which we are observing and where we are living in.

The space dimension's momentum cancellation and origin of the matter:

Every object of the Universe is suspended in space and can travel in six different directions! – East, West, North, South, Up, and Down! All those six different directions are representing the six single space dimensions of the space of our Universe! Our space cannot be three dimensional it is six-dimensional!

In this configuration, we are observing three pairs of single dimensions with opposite direction! - This is very important because this configuration provides the required conditions for the cancelation of the dynamic momentum of the space dimensions and as a result of this, they are forming normal steady, define volume of space, where itch space dimension preserve the spatial orientation of its energy field.

(For example - It is similar to the scenario where two identical spinning marbles collide and are canceling their kinetic energy, but still preserving their rotational spin). From the nature of the Time Dimension, we know that every single dimension is physical and carries energetic dynamic momentum! The opposing direction of the space dimensions providing a condition for cancellation of their dynamic momentum. Understanding the space dimensional momentum cancellation has enormous implications because exactly this phenomenon is explaining the origin of all the matter of our Universe!

A well-known fact is that cancelling of energy fields produces heat and releases energy, which effectively is mass! The observed fixed proportion ratio of matter to space volume is very solid evidence in support of this explanation - (because the universe is homogeneous in each direction and distance!) This is logical explanation of the observed properties of the Universe, where everything fits perfectly to this explanation without violating the law of physics, or necessity to be created mystical substances in order to prop up unworkable models and cover-up incorrect fundamental assumptions which are violating the law of Physics!

Some could ask do I have any evidence in support of my theory? The answer is yes! - There is not just "any" evidence, but there are millions of evidence all around us, we just have to know where to look for them.

I will give you some examples and will start with Time - we know that Time is a single dimension and we know that we have only one Time. Nobody asking proof for the existence of time, and nobody suggests that there is more than one time! That's why we have only one watch on our wrist. If we have more

Time dimensions, we will know this and will need to have more watches on our wrist. - It is the same situation with space dimensions (or directions). Same as our knowledge of Time, we know that there are six space directions, and I don't have to prove this! The fact that we can travel in six different directions from each reference point of our space, is undeniable proof for the existence of six single dimensions of our space. If there are three dimensions only, we will have no clue for the existence of six space directions - we will know only three directions! We encounter only what is included in our physical World, no more, no less! It is a simple fact that - we couldn't travel in six directions if our space is three dimensional only! Full stop! This is my answer to the question - do I have any proof for the existence of six space dimensions. The proof for it is printed on every map, every compass, and GPS, where it is clearly visible the four separate surface directions – East, West, North, and South. The existence of the other two directions - Up and Down, everybody knows them, and I don't have to prove their existence. - (To define any point on a surface, you need four vectors, but to define and to form space, you need minimum six vectors, which represent the six space directions!) space has no special reference point in it. – Every point of space is reference point – (HERE and NOW), where from our perspective the Time and Space dimensions are crossing and the six dimensions are beginning in each direction. Our unique personal point of existence (no matter where we are situated) always providing us with the ability to travel in each six space directions!

The next crucial point of understanding Space is its relation to Time! The assumption of the current 'Standard Model' that Space and Time are physically incorporated in one substance called - 'Spacetime' and forms four-dimensional space is absolutely incorrect! Einstein's 'spacetime' is physical incorporation of four single dimensions, which is a real physical mess! It is not working and cannot produce, or explain anything! An additional active dynamic dimension as (Time) incorporated in the well proportional and steady space will create a fatal distortion of any shape and any object! Space and Time are separate physical entities and the relation between them is different than the currently assumed 'Standard Model'! Space and Time are not physically incorporated! They just share the same volume! **Time is dynamic, but Space is steady, they cannot be physically incorporated!** The six physically incorporated space dimensions are cancelling their progressive, dynamic momentum and forming just a steady volume of space and a defined amount of matter! There is no such thing as 'spacetime' – there is just time in space and space in time! To understand the difference between those two terms is of fundamental importance for the correct understanding of the physical order and principles of our Universe!

Some could ask where the antimatter in my model is? My answer is that this is purely a speculative question unrelated to my model. First, the appearance of antimatter in the particle accelerators is irrelevant to the conditions of energy cancellation of space dimensions. And second, this question could be raised only, if it is a performed experiment with the cancellation of the energy of two opposite dimensions and the antimatter is the result of this experiment. - Most likely will be that the cancellation of space dimensional dynamic momentum will produce neutrinos and antineutrinos. So far, currently, the neutrinos are known as the smallest and the most abundant particles in the Universe. The neutrinos and antineutrinos are not annihilating each other and are able to change the "flavor" (or change into each other). - This could explain the hypothetically assumed "lack of antimatter" in the Universe!

What is Energy?

We are using and defined many forms of energy in our everyday life – as Kinetic energy, Potential energy, Thermal energy, Electromagnetic energy, Strong and Weak Nuclear energy, Pressure, Magnetic and named... We know that the fundamental form of Energy must be in only one form and to apply to all known different manifestations of Energy. Unfortunately, we are living in total lack of understanding of the fundamental elements of the World.

Current science cannot defy even the basic components of Universe: as Space, Time, Matter, Information, Consciousness, Law of Physics needer their origin. Same situation we have with defying and understanding of Energy. – Modern science has no answer to this question.

I will try to change this situation and explain what actually is the basic – the fundamental form of all known to us forms of Energy. I will start directly with the definition of what the Energy is and then I will give a detail explanation: - The Energy is a disturbance of the equilibrium of Space.

A few pages above we have discussed the structure of Space. For simplicity and to be easy to understand the contest of Energy, I am providing a graph with the micro-layers of Space dimensions below.

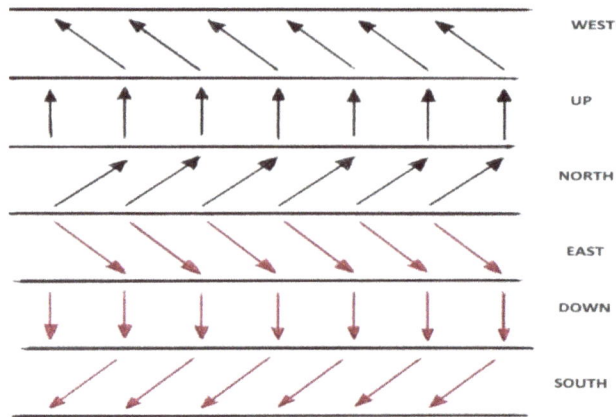

We can consider Space as a balanced energy field. Any force inserted there, which will push the boundary border between two layers of Space and will create energy vacuum, or energy pressure in the point of interference. The Law of Physics dictates, that the equilibrium must be restored. <u>The tendency of the Physical system to restore the Equilibrium is the manifestation of Energy!</u>

I will give one example: If electromagnetic wave is travelling in one of the layers of space, the pulsating sinusoid of the electromagnetic wave will expand or contract the boundary of the layer. This interference with the shape and equilibrium of Space is the basic – it is the fundamental form of all kinds of Energy. (See also the graph on page 174)

Primordial space; Origin of the six-dimensional space and its energy field.

From the chapter above, we reach an understanding that the origin of all energy of our Universe is coming from disturbances of the equilibrium of the six-dimensional space. This realization is bringing us close to the primordial source of the Universe - which is the single and ultimate source of everything around us.

It is obvious that our six-dimensional space is a composite structure designed to serve its purpose.

Every composite structure has its basic element, which is used in this structure to be constructed. For example, a car engine is constructed of a single substance – Metal! The question is – what is the single substance for the construction of our Space?

We have a few clues which are leading us to the single basic element of the space dimensions. The main clue is that all six dimensions have the same physical property - they just have a different spatial orientation.

161

When we eliminate their differences we will end with a neutral spatial medium, having a total lack of orientation or any active physical properties. - I will call this medium "Primordial Space". The lack of physical "tension" and any physical interactions means that the Primordial Space is not containing energy. - (I am not commenting on who or how our six-dimensional space has been created). I am concentrating only on the Physical mechanism of the structures and the resulting property of the building block of space. From the diagram above we can have a visual picture of how the six space dimensions co-exist in confronting equilibrium, where the spatial orientation is creating a conflicting physical property.

It becomes obvious that the lack of physical property and lack of energy of the Primordial Medium coming from the absolute uniformity of its structure. The key factor in our six-dimensional space is in the physical confrontation of the differently oriented space dimensions. - Or we can say that the energy source of our six-dimensional space is coming from their different space orientations! This is a very simple configuration, but as usual, the most genius inventions are the simplest ones. We are using this method every day in the construction of our composite materials and know that from uniform structures, we can produce different materials with completely different properties. This is how our Space has been constructed.

GRAVITY

The inherited incorrect - expanding and open space model of The Big Bang theory and Einstein's weird assumption for space curvature create a real physical mess and are stripping us of the chance to understand the real nature of Gravity and what the attractive forces are. Those fundamentally incorrect configurations are leading to the creation of endless weird mathematical models and theories, which do not represent reality, not make sense and are not explaining anything – as gravitons, gluons and the "discovery" of Higgs boson! In reality, there is nothing mysterious or complicated to understand how the attractive forces and gravity works. It is not very difficult to comprehend if we keep our feet on solid ground and using the law of Physics, proven facts and observations to build a correct model and correct configuration of the fundamental building blocks of our Universe:
We have to realize, that in our physical world of particles and waves, there does not exist any substance, force, or mechanism, which can provide physical attraction! We can push and propel objects with particles, but we cannot pull them back or attract them with particles! – It is impossible! So... how with

propulsion only in our arsenal, we can have attractive forces?

I will start with the law of physics which states: that energy and matter are moving from a state of higher energy level towards a lower energy level. – This is the key behind the existence of the attractive mechanism of all known physical forces in our Universe!

The forces of Gravity and Electromagnetism are acting in the same way! In the diagram below it is visible that the energy field surrounding the two objects is uniform in all directions and acts as energetic pressure on the objects. The energy field is canceled in the space between them, and this creates an energetic imbalance- (as energy vacuum). As I stated above, the law of physics dictates, that the outer energy (pressure) will push the two objects together! – This is what we are observing with magnetism, <u>and this is the actual Universal Principle of Attraction!</u> – There is no "Mystery" involved. This is a simple physical process and is the common mechanism for all the known attractive forces in the Universe! The science is assuming that the vacuum (or space) is filled with energy, - a lot of energy! (The explanation is in 'Understanding Space') There is a continuous argument for the strength of this space or (vacuum energy) where the predictions vary between 10^{-9} to 10^{113} joules per cubic meter. The last figure is enormous! Some scientists are stating that in one cubic centimeter of space, there is stored more energy than in the entire Universe! Even if just a fraction of this claim is correct, the closed physical configuration of The Universe will provide conditions for this energy field of space to acts as a uniform energetic pressure surrounding all material objects and act as a pivoting point for the attractive forces. The energy disturbance or energy cancellation and imbalance between any two material objects will deplete slightly the density of the space energy field between those two bodies and the energy imbalance will start pushing those two bodies toward each other, similar as the two objects with opposite magnetic charges shown on the diagram below.

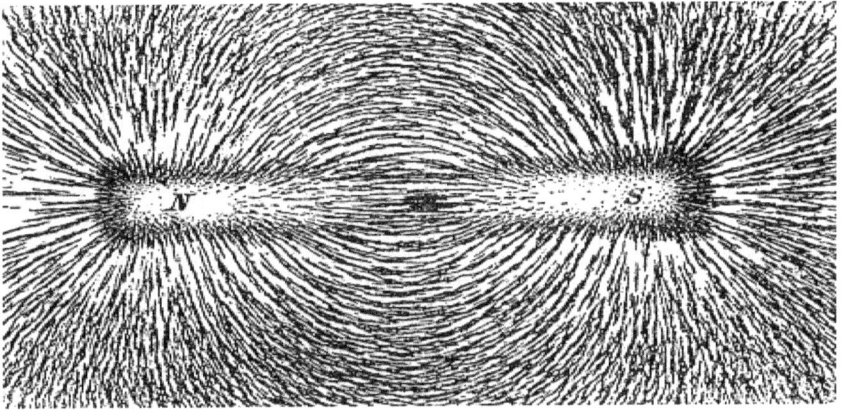

The opposite charges are cancelling the energy field between them and producing "energy vacuum"

This is a well known physical process where the natural tendency is to move from a higher energy field towards a lower energy level. This is the mechanism, which explains the "mystery" of gravity, and all attractive forces in the Universe! The electromagnetism and nuclear forces are working in the same way – they are a disturbance (or local imbalance) and cancellation of the energy field between two bodies!

It becomes very easy to understand how force imbalances are producing attraction! This is the answer to the question of what is the nature of the attractive forces, what is Gravity and how the Gravity really works! The opposite rotation of the Birkeland current filaments and the opposite direction of the induced electrical currents to the direction of the primary source are supporting my claim for their real origin. –

(The necessity and the tendency of the energy disturbance and imbalance of space energy field are to be restored.

Understanding the atomic nucleus structure, and the union of Quantum mechanics and Newtonian physics

The "Standard Model" gives us a very illogical and unworkable model of the atomic structure. To hold together their nucleus, mainstream scientists have created hypothetical mass-less particles - Gravitons and Gluons which are supposed to act as nuclear glue. This hopeless invention wasn't able to explain anything, and has dictated creations of more mysterious substances and mechanisms, in the form of "God particles", but even "God" wasn't able to fix this mess, because the proposed model suffers a range of conflicting properties and fatal configuration flaws. The top echelon of physicists doesn't understand such simple things as the fact that nobody, - (even "God") cannot produce attraction by using particles only! Despite this fact, the top guys claim that they can do exactly this! - (I am sorry, but you cannot produce rope from loosed sand)! Full Stop!

The current official model of the atomic structure is unnecessarily complicated with the invented non-existing particles and forces. This is creating a conflicting scenario, and an unworkable model, where the atomic elements and forces are in direct conflict with each other. To get reed of this mess I will start with one simple fact – Matter and antimatter are annihilating each other when coming in contact. This simple fact is telling us that the atomic structure of matter and antimatter is held together by electromagnetic force because the only difference between matter and antimatter is their opposite electromagnetic polarity. The annihilation is no more or less a simple electrical ("Short Circuit"). This "Simple Fact" is telling us that we have to concentrate on this undeniable fact! – How is the configuration of the electromagnetic forces and polarity in the atomic structures to be able to hold together all its elements?

If we do this, we will start from the correct position – from the fundament and go up and build a correct model of atomic structure!

In the official model, the attractive particles "Gluons" are in proportional dependency to the nuclear particles, and are evenly distributed (exchanged) between them. This configuration cannot explain how and why the mass deficit of the atomic nucleus is producing nuclear fusion. This configuration will allow the atomic nucleolus to be stable and to grow exponentially to any size, even to become bigger than the Universe! – (This crucial fact has been ignored). When we start using logic, it will become obvious that the strong nuclear (attractive) force must be allocated only in the centre of the atomic nucleolus, where the energy of the nuclear constituencies opposing and cancelling the space energy between them. The result is that the centre of the nucleus becomes the point of partial cancellation of the energy field of space, which acts as an attractive centre with a range of nucleus only. - This explains why the strong nuclear force has such short range and why is the mass defect of the nucleus - (the missing binding energy) it is undeniable proof for it. This explains why the atomic nucleus cannot grow exponentially and become unstable beyond the 82nd element! The spontaneous nuclear decay of the lighter elements is a clear sign that the atomic nucleons are slowly gaining their missing "binding energy", which in effect is gaining of energy from the surrounding energy fields. And this small and gradual gain of the "missing nuclear energy" is clear proof of the energy cancellation nature of the attractive nuclear force, and that the actual source of the attractive nuclear forces, gravity and electromagnetism has the same origin and same attractive mechanism, which is a local cancellation of the space energy field! It's pointing out also, that the "Weak nuclear force" does not

exist! The structure of the atom is based on groups of <u>six</u> fundamental elements: - There are <u>six</u> stable electron shells. The proposed or "detected" <u>six</u> Quarks are named: Up, Down, Charm, Strange, Top, and Bottom. The known leptons are also <u>six</u>. The existence of <u>six</u> fundamental elementary particles with <u>six</u> specific "angular momentum" cannot be accidental, and this specific number is confirming the existence of <u>six</u> separate space dimensions, which are the actual instrument to separate those particles by giving them <u>six</u> different space orientations, which the top echelon of science is naming as "angular momentum" (or spin). If the top scientists have known earlier, that their "angular momentum" is just ordinary <u>six-dimensional</u> space orientation, definitely, they have named the Quarks as: East, West, North, South, Up, and Down. It is obvious that the variety of elementary particles can differentiate and be separated from each other by their different <u>six-dimensional</u> space orientations. It cannot be co-incidental also the fact that the stable electrons shells are <u>six</u>. It becomes obvious how the variety of elementary particles is differentiating and separated physically from each other because of their specific different <u>six</u>-dimensional space orientation, (or space dimensional separation). The neutrinos are the most abandoned particles in the Universe. They are also the smallest particles currently known. The physical properties of neutrinos are one of the most compelling evidence for the existence of <u>six</u> dimensions and <u>six</u>-dimensional space orientation because the neutrino number is also <u>six</u> – (electron; muons; tau, and their three antiparticles). I quoted them as **six**; because the fact is that the neutrinos and antineutrinos are not interacting with each other, and not annihilate each other. Actually, they are the same particles, just having different (opposite) space orientation, which corresponding to the three pairs of oppositely oriented space dimensions - (east and west; north and south, and up and down)! The neutrinos are the lighter and the fastest travelling particles. They are not interacting with anything, and not annihilate each other. This physical property makes the neutrinos a good candidate for the first building block of matter. In such a case they could be instrumental for the nuclear forces too. The electrons decay in quarks, and we don't have to ignore also the possibility that the quarks could be composed of neutrinos. The fact, that we cannot detect any internal quark structure is no evidence of lack of neutrinos there because we don't have any method for direct detection of neutrinos. The quarks have different weights and different physical properties. Four of the quarks are not stable. The different physical properties of quarks are telling us that they cannot be the prime building block of matter, but are composite particles with different quantity of energy, matter, and particles involved in their structure.

Let's see what the impact of the six-dimensional configuration of space will be on our understanding of the property and structure of matter. A well-known fact is that the Quantum mechanics and the Newtonian physics of the macro-world are incompatible, despite the fact that the biggest and smallest structures are situated in the same space, experiencing the same physical conditions and are ruled by the same laws of physics. This fact leading to the realization that it's not the law of physics which are different, but the size of structures making the difference. - The small size of particles is giving them the ability to "slide" between the layers of space fabric. The current understanding that identical elementary particles cannot occupy the same space unless possessing different spin and angular momentum is leading to the realization, that exactly these functions are instrumental in differentiating the identity, functions and spatial separation of the elementary particles. Despite such understanding, the 'standard model' is foggy and mysterious, because is not providing a correct understanding of space structure and mechanisms, which are able to explain the weird particle properties and behavior. Einstein's three-dimensional space is helpless to accommodate these properties and functions and to give us the necessary explanation. The correct understanding of what actually a single dimension is and understanding that space is six-dimensional is shed light on this puzzle and explains much better the weirdness of Quantum mechanics and the strange behavior of the elementary particles. A Logical conclusion is that the fine layers of space are providing the conditions for the elementary particles to use different oriented "dimensional layers of space" in order not to occupy the same space. I will explain what that means: Imagine that the space dimensions are not mixed as one homogeneous fluid, but they are tightly packed as pages of a book. The fine and tiny structure of space is acting as homogeneous and uniform space for the large objects, but the smallest elementary particles can be on different pages or hide between the pages. The space layers orientation is putting the particles in different "Phases" (Similar to graph of three phase AC).

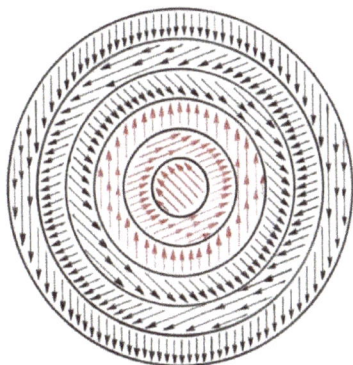

This is the fine space layers of the atomic structure

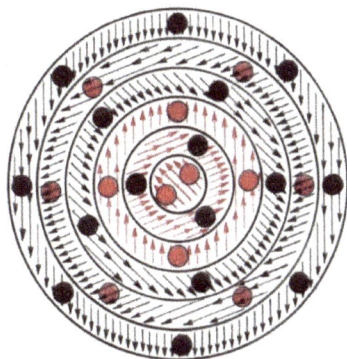

Shell configuration of atomic nucleus where neutrons are separating the proton shells

Atomic structure is similar to the layers of onion structure – the space layers are wrapped around the nucleus because space is an energy field and the created energy "vacuum" in the centre of the atomic nucleus inevitably will create a few spherical energy levels around the nuclear centre. The centrally "cut off projection" (see above) of the atomic structure is similar to the rings of a tree trunk. The different rings represent the different dimensions, and the distance between them provides room for the elementary particles to be in one, or be between two single space dimensions, allow them to jump from one to another level, or spread and be difficult to locate them.

The evidence for the shell configuration of the nuclear structure is coming from the fact that elements with up to three electron shells (the number of nucleons and electrons are related) having (nuclear mass deficit) - releasing energy in nuclear fusion. Elements with more than three electron shells having (nuclear mass excess) - or need additional energy to be able to fuse their atoms, (Because the next three shells are situated in dimensions having opposite directional oriented energy field to the first three dimensions). Further proof for the shell configuration of the nucleus is the fact where when the nucleus is bombarded with electrons, some elements first emit neutrons, but some elements prefer to emit protons. This can be explained only with shell configuration of the atomic nucleus, where the different shells are constructed respectively with an excess of neutrons or protons. The configuration of the nuclear forces is leading to the conclusion, that the protons are separated by neutrons where in the first, third, and fifth nuclear shells the protons are dominant, but in the second, fourth and the sixth shell, the neutrons have the majority. The currently unexplained nuclear forces easily will be explained by the six-dimensional space configuration of the atomic structure. The space energy cancellation is located in the centre of the nucleus and the space energy distortion (deficit) is just in centre of the atomic nucleus - (It is confined energy between the nucleons). This is the reason the strong nuclear force to have such a short-range and not to leak out. This configuration is explaining also how the elementary particles can be in superposition, and why the atoms are stable, why electrons are not losing their energy and are not falling and crashing in the nucleus. This configuration is explaining also how the different space orientation and space position of the elementary particles is separated and make them different from each other, but not stopping their physical interactions. As soon as the material structure grows above the thickness of the "Pages" or (the fibber and layers) of Space Dimensions, Space starts acting as a uniform homogeneous space for the largest objects, and the Newtonian laws are taking over. This explains the weirdness of Quantum Mechanics and why Quantum Mechanics cannot rule the biggest structures and is valid only for the microscopic world. – This is the currently missing unification point between Newtonian Physics and Quantum Mechanics.

The "solid" state of particles is also based on the same principle – It is a tight and well balanced concentric configuration of standing waves around the centre of space energy cancellation point.

Or in short, there is no such thing as solid particles or solid matter! – The solid matter is just clever configuration of well-balanced opposing forces! The proof for this claim is the fact, that when matter and antimatter get in contact, they are annihilating each other and are transferring in pure energy!

Spontaneous nuclear decay and the Weak Nuclear Force

From the chapter above, we reach an understanding that the nuclear binding energy is the energy field cancellation between the nucleons. Also, the known nuclear mass deficit is giving us understanding, that the opposing energy of the nucleons is providing the space energy cancellation between them, which is the strong nuclear force. This knowledge is giving us a good understanding of the structure of the atomic nucleus, where the energy cancellation is in the centre of the nucleus. This nuclear structure making the outer shells of nucleons very weak attached, and further growth become impossible. In the heaviest elements beyond Led, we observe spontaneous nuclear decay. For an easy understanding of the concept of spontaneous nuclear decay, I will use the example from thermodynamics, where the individual atoms and molecules of matter are having slightly different energy levels from each other. The same phenomenon is valid for the nuclear energy level of different atoms. The overall decay rate for specific elements is uniform, but the life of a specific atom is dependent on the energy level of the atomic nucleons! - This is the real explanation for the reason for the spontaneous nuclear decay. - The outer shells of nucleons are losing at very slow rate its (excess) of binding energy, due to interactions with the surrounding energy fields. And when the specific atom reaches the lower limit of binding energy, the atom decays.

The creation of the hypothetical "Weak nuclear force" is a result of the total miss-understanding of the nature of the attractive mechanism of the physical forces. Nobody ever has detected this invented "Weak" force. There is no single experimental observation of this hypothetical invented force. It is more than obvious, that the "Weak nuclear force" does not exist!

UNCERTAINTY, SUPERPOSITION, AND VELOCITY OF PARTICLES

We know a load about these phenomena's and scientists have done millions of different clever experiments to find the mechanism and the reason for the "Weirdness" of Quantum Mechanics and the unexplained "Strange" behavior of the elementary particles. So far, the explanation of this puzzle remains elusive and the solution is distant as one hundred years ago when Einstein and Bohr have clashed and fought on this subject.

The "Theory of Everything" within the boundary of the law of Physics is able to explain all current uncertainties and puzzles in Physics and Astronomy simultaneously.

 Probably, will be interesting to you to learn how the Theory applies and solving the biggest puzzles in Quantum Mechanics - (The Uncertainty).

To be able to understand the core of these phenomena's we first have to come in clear them with our understanding of what is defining the position and what is defining the velocity of the objects.

Position: the best description of the position of some object is characterizing as - Here and Now or There and Now. It is obvious that in order to define the position of some object, we have to know simultaneously its position in Space at the exact Time.

Velocity: velocity of some object is the Time difference between two positions in space - (Here) and (There). This means that the phenomena of velocity involve two different places in Space and two different values of Time.

Uncertainty: from the "Theory of Everything" we know that space is formed by tiny layers of the six space dimensions. This space structure is allowing the wave state of particles to exist in one, or more space dimensions. The saturation of particle waves in different dimensions can vary. This is the main reason and mechanism for the observed "uncertainty principle" in Quantum Physics - (spreading the particle waves into a few space dimensions).

Consciousness, Reality and Collapsing of Wave Functions

The Universe is constructed from physical and non-physical components. The physical components are Space, Time, and matter. The non-physical components are Consciousness, the Law of Physics, and Quantum Information. The non-physical components are made of a currently unknown form of energy.

Until now, there are two main dominant approaches in explaining the World - Materialism, and Spiritualism.

Both philosophies are primitive because they are considering only half of the real property of the Universe.

Currently, Quantum mechanics is considering only the material aspects. This primitive approach is the reason why Quantum mechanics cannot provide a credible explanation of the World.

To be able to understand the World, we have to use a Scientific approach. The scientific approach considers all facts without prejudice. We will use only the Scientific method in our consideration:

The Universe is constructed as one giant super-computer, where the consciousness is corresponding to "Window Software." The Law of Physics is like our "RAM" (operating memory), and the Quantum Information is the information produced from the computer programs and functions.

The fact that all living organisms possess a consciousness and most of them not have nervous system can explain to us that our consciousness is not a product of our mind, but is part of the "System of Universe."

Our brain is just our processor, which is using consciousness as an operational functional program.

The knowledge that the consciousness of the Universe is its functional "Software", which commands and regulates all physical processes in the Universe is giving us an understanding of relations between consciousness and the quantum information.

And now back to our subject: Why our consciousness is collapsing the wave functions of the elementary particles? It becomes clear that our consciousness is not insulated phenomena, but is part of the system, where the information is controlled by the consciousness. Having this understanding is easy to realize, that when our consciousness is interacting with the delicate state of the particle, it is affecting its information. The interaction of our consciousness with the particle information is changing its physical state and is collapsing its wave functions. - Is this just random interaction, or are pre-determined phenomena currently is known. One clue is that there are a few cleverly inserted "limits of knowledge," which is preventing us from gaining knowledge of how to create matter and how to predict and change the physical property of the Universe.

I am suspecting that such cleverly inserted key limits of knowledge cannot be accidental.

On the claim: that the "observer creates the reality," I will present short contra-argument: If the observer creates the reality, then who creates the observer and where the observer is situated?

LIGHT

Light is the unifying link between thermal, kinetic energy and electromagnetic property of The Universe.

The Light is coming in two forms - as particles - (photons) and as (Electromagnetic waves). The confusion in the accepted model for the property of light in the "Standard Model" of Physics is coming from the fact that Photons are not composite particles. The officially accepted view is that Photons are mass-less. We have to come down to reality and to realize that "Something" cannot be made from "Nothing" and if Photons exist in form of particles capable to apply pressure on the illuminated surface, this "Pressure" is an indication of the mass of the coming Photons. - Photons have momentum, and momentum is the combination of mass and kinetic energy! We also observed, "Gravitational Light Red-shift" which means only one thing! - Gravity pulling and depleting the energy of the photons. Gravitational Lansing also proving that gravitational attraction is altering the path of the light. If Photons are mass-less the gravity will not have any effect on them! We are considering Photons as fundamental particles because Photons are not composite particles and the photons can disappear without a trace - something, which the composite particles cannot do.

So what actually is the nature of Light?

I will start the explanation with the wave function of the light: The simple definition which I can provide is that the Light is a travelling wave through the medium of Space.

This definition is not too far from the officially accepted, but in reality it is fundamentally different. The difference is coming from the involvement of Space Medium. In mainstream science mentioning of "Space Medium" is Taboo! - Again we have tho come down to reality and to realize, that there ar no such things as "Wave Travelling in Nothing". The basic nature of wave is "Harmonic Disturbance of Medium".

I am providing the graph where in simple way is represented what is the interactions between Light wave and the Space medium. This is the fundamental diagram, explaining the interactions between Space and Waves. - (See the graph below)

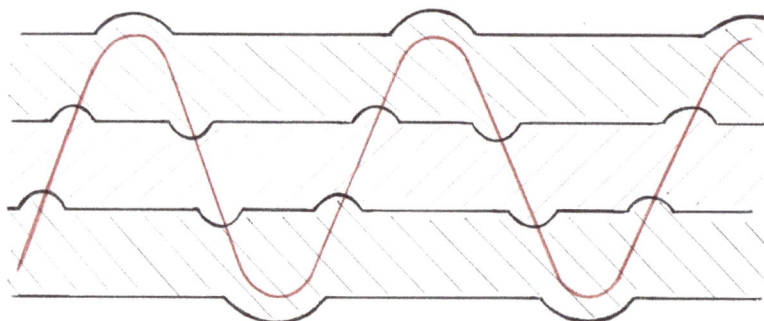

This is how the travelling waves create distortion of the layers and boundarie of space

The travelling light wave actually is curving the boundaries of the space dimensions and creating pulsating local energy excess or energy deficit in the points of interactions with the Space Medium.

The energy of the light wave is corresponding to its frequency. It is easy to understand that the higher frequency will have a bigger amplitude, and in a certain length of space, will deposit more and stronger disturbances of the equilibrium of Space Medium.

Now, on the solid state of Light - Photons: In the previous chapter, we have defined that the solid-state of elementary particle is point of space energy deficit surrounded by a standing wave. This also is the configuration of the "Solid State of Photons". Photon is just an ordinary fundamental particle.

On the question of how the electromagnetic field has direction and spin (the right-hand law) the answer is that the photons have direction and spin and are interacting with multi-directional medium (Space). - In such dynamic interactions, the normal outcome is that the result also will have direction and spin.

How works the mechanism of transformation of spread travelling wave into the concentrated form of particle, the current science still has no answer. It is a very complicated process to be understood because there are involved many different factors, which our science is not willing to address yet. - The Law of Physics, Consciousness, Information, Space dimensions and orientation. All these elements are involved together in quantizing and defying particles and waves. I wish to be able to explain the mechanism of wave functions collapse a bit better, but at least I have the courage to revile to you which phenomena are involved in the formation and functioning of the Physical order of the Universe.

THE FIRST BUILDING BLOCK OF UNIVERSE

In the beginning, I have defined the six building blocks of the Universe - which are Space, Time, Matter and Consciousness, Law of Physics and Quantum Information. I have stated that in order to have a correct understanding we have to consider all the elements of the Universe without prejudice.

We have managed to get deeper into the material aspect of the Universe and to find the most fundamental blocks, which are forming the material aspect of the sophisticated structure of the Universe. So far we didn't finish the job, because in our picture remain the first building substance of the material structures and the three non-physical components of the Universe. – Consciousness, the law of Physics and Quantum information. We know that the Law of Physics and the Information can be a product only of an intelligent phenomenon – (consciousness). We will live them aside because they are in the category of composite structures, which are a by-product of the other elements.

And now we have ended with the last two fundamental components of the Universe – The Primordial Medium and Consciousness. The Primordial Medium is a smooth and passive substance and does not possess any physical property or creative power. - In contrast with this lifeless property of the Primordial Medium, the Consciousness possesses many welknown interesting phenomena and creative element. That means that Consciousness is a composite structure, capable of creative and intelligent actions.

We know some of the property of our own consciousness: We possess love, compassion, dedication, curiosity, loyalty, humor - we enjoy music, poesy; admire beauty, desire for adventures. All this quality is not coming and belongs to the physical aspect of the Universe! They are a product of The Consciousness! The Universe is constructed on the base of two fundamental elements only – The Primordial Medium and Consciousness.

The material structures of the Universe are constructed only of the structures of matter, but the living organisms are a combination of mater and consciousness. The fact that the Universe is constructed in the same way is giving us the understanding that we can put the universe into the category of the living organisms – One enormous and sophisticated living structure capable of producing life!

MATTER/ ENERGY EXCHANGE AND THE RECYCLING MECHANISM OF THE UNIVERSE:

The ancient philosophers have achieved great understandings of The Universe and structure of matter. Their scientific data has been very limited, but by using intelligence and logic they have been able to understand the Heliocentric Solar System and by using the size of the shadows of the solar and moon eclipses to calculate the distances and sizes of the Sun, Moon, and Earth. By using observations of chemical reactions, they constructed a very modern atomic model of matter. All those amazing achievements the ancient philosophers succeed to made because they used logical analysis of the available facts!

We currently possess an enormous amount of scientific facts and data, but in general, we are far behind our ancestors in the understanding of the World. - Why?

The answer is simple; we have departed from common sense and logic in every aspect of our life, especially in science! The mainstream science considers only ordinary matter and gravity to explain The Universe, but the problem is there, where 99% of the Universe is in the form of plasma, and plasma is ruled by electromagnetism, which is many billion times stronger than gravity! By considering only 1% of the existing matter and ignoring the strongest and dominant electromagnetic force, the present science will never be able even to get close to an understanding of how the Universe works! - This is not a very pleasant situation, and we have to address it!

To be able to construct a realistic model of the Universe, we have to include all known existing elements, without preferences and ignorance!

In the interstellar space, there is spread a huge amount of defused matter in the form of clouds of dust, plasma, gases, ionized particles, molecules, and neutral atomic matter. (See the image on page 2)

Despite the lack of density, this diffused matter is representing most of the matter of the Universe, because the distances between the stars and galaxies are enormous. The other phenomenon, which is not considered by the current model, is the electromagnetic property and interactions of the celestial structures. Everybody knows that the Earth has a magnetic field that keeps a compass needle facing north-south and protects us from harmful solar and cosmic radiation. Every star and most of the planets also have magnetic fields. Each galaxy and each structure of the Universe also contains magnetic fields. The Universe is a dynamic place, where everything is moving, shifting and rotating. The Earth's magnetic field is rotating in the Solar magnetic field. The Solar magnetic field is rotating in the Galactic magnetic field. The Galaxy is rotating in the magnetic field of the Universe. We know that when magnets or magnetic fields are rotating in another magnetic field this rotation produces current – (the dynamo effect)! The intergalactic current produces enormous magnetic fields! Electromagnetism has a bonding effect, similar to gravity and also has unlimited range! The electromagnetism is energy, which is acting also as mass! - The current science is not taking this into account. The electromagnetic forces are many billion times stronger than gravity - (10^{38}). Some electromagnetic processes in the Universe possess enormous strength! They are able to produce up to 40 million times more energetic beams than our best particle accelerator! The magnetic field of a quasar reaches 200million times the strength of the earth's magnetic field! The dynamic electromagnetic interaction of moving and rotating structures of Universe is the main mechanism behind the distribution of matter and energy between galaxies and stars. Electromagnetism is eroding the dying stars and they are releasing enormous amounts of charged particles back into space, where the electromagnetic currents ferry and distribute plasma and diffused matter to the regions where the new stars are forming. The streams of plasma are stretching and connecting the Universe structures and its' movement is dragging the neutral particles and these combined streams are feeding the galaxies with the necessary matter for the new stars formations. Those streams of diffused matter have been invisible in the past, but this is not the case for our new sophisticated instruments. - (See the graph on page 2)

The work of the Universe is based on very simple, but very reliable systems – balanced dynamic and continuous energy - matter exchange. To understand how the system works, we have to understand how the structure of the Universe is designed. The fundamental elements of Universe structures are: Space, Matter, Time, and the Laws of physics. We mentioned above, that for every amount of matter there is an allocated specific amount of space! That means, that the Universe is acting as a closed physical system, (regardless of it is endless or finite), where space has constant volume, and there is an inserted specific (proportional) amount of energy (matter), and this inserted energy has nowhere else to go! Our observations show that hydrogen is the most abundant element in the Universe. The hydrogen is the most energetic element of all elements! When the hydrogen starts fusing into heavier elements the nuclear fusion releases enormous amounts of energy into the space of the Universe! This released energy boosts and increases the rate of the reverse process - which is the electromagnetic fission of heavier elements. This reverse process is restoring the balance ratio of hydrogen/heavier elements; the amount of space diffused matter; and saturation of interstellar space with thermal and electromagnetic energy.

The Universe is possessing enormous systems of dynamic plasma distribution, which provides material for continuous matter/energy exchange between all celestial structures. The magnetic field and the gravitational motions of celestial bodies are producing enormous electromagnetic fields and currents, which is forming powerful electromagnetic "vortexes" – (which are mistaken to be Black Holes). There the released energy is breaking the elements back into hydrogen! The unexplained abundance of neutrinos coming from all directions is the actual proof of the continuous recycling processes of the Universe. -

And this exactly is the self-balancing mechanism of the Universe, - the finetuned ratio between the released energy of hydrogen fusion and the quantity of heavier elements! - As soon as the amount of heavier elements gets increased, the amount of interstellar electromagnetic energy is increasing with geometric proportion and boosts the reverse process of electromagnetic nuclear fission which is breaking the heaviest elements back into hydrogen and stabilizing the precise chosen balance-ratio of the three components: - heaviest elements to hydrogen, and the saturation of interstellar space with thermal and electromagnetic energy. - And this exactly is the picture, which we are observing: One vast, incredible and beautiful Universe, balanced with absolute precision, stability and simplicity, where the fine balance of forces and kinetic energy provides continuity for the eternal dynamic existence of Universe structures!

The electromagnetism is responsible for one more so-called "Puzzle" – the rotational unity of the galaxies!

The electromagnetic field of the galaxy not only acts as an additional gravitational unifying bond but also produces static electrical polarity and bonds between the molecules and neutral particles, which provides additional bonds and solidifies the galactic structures as one rigid body with uniform rotation! – This is a simple physical phenomenon, which explains the rotational unity of the galaxies and makes unnecessary the invention of the mysterious "Dark Matter" theory to explain this simple physical interaction! Our science stubbornly is obsessed with the idea to find and use nuclear fusion but the endless experiments and billions of dollars spent, producing negligible results! It is more than obvious, that this approach is fundamentally incorrect and not working! It would be beneficial if science starts to look to find the real mechanism - where and how exactly the Universe is fusing and tearing apart the heavier elements back to their prime form of hydrogen! It would be good to pay attention and study carefully the composition of falling matter into the electromagnetic "vortexes" of The Universe and compare it with the composition of the plasma jets coming out of those "vortexes". The difference in compositions will tell us what actually is going on in those ultimate milling machines of The Universe! The process of spontaneous nuclear decay is not considered in nuclear fusion research – Why? If we are looking for fusion, would be good to understand the opposite process too. The nuclear decay is an obvious sign of loss or gain of nuclear bonding energy, and this "lost" energy could lead us toward mastering the elusive nuclear and gravitational forces! I believe the answer is in not the pressure and temperature alone, but when they are in combination with the strongest force of electromagnetism - together they will be able to fuse, or break the elements nucleolus easily! Such places emitting energetic particle jets we are observing in the Galactic centre, Quasars, Pulsars and nebulas.

Our Science has to start using common sense and logic and start looking to find the source and the recycling mechanisms of the Universe. We have to find where and how exactly the Universe is fusing the lighter elements and breaking the heavier elements back to their prime form of hydrogen - the mechanism is out there and our job is to look for it and to find it!

Someone could ask me why are no mathematical formulas in my Theory? The answer is simple: the mathematical formulas have no meaning! The scientists have tried for millenniums to explain the world with mathematics, but they didn't succeed. The Universe is built with logic, harmony, and purpose. Ask yourself: what we see, what we hear, what we fill and what we know about the world around us? - We see the beauty of nature in colors. We hear the sound and the harmony of music; but why the colors are seven? Why the notes are also seven? What makes the colors and notes to come in harmony with Nature? We are part of the Universe and part of Nature! We are filling the World exactly how it is! We know that there are six world directions; we know also that Time exists! - Space and Time are the Seven Dimensions of the World. The sound and light are vibrations, and only when they are in phase with the Seven Space Dimensions, only then they become in harmony with Nature. That's why we are enjoying the harmony of music and the beauty of Nature! With our eyes, our fillings and our senses we can understand the World much better than any mathematical formulas! Mathematics is not providing explanations! If you want to understand the World, live the mathematic aside and open your eyes and fillings! The World is out there!

Currently, we have enough data to be able to draw a correct and realistic picture of our Universe!

How I expected at the beginning of this chapter, The Universe was able to surprise us again, and to increase its size in parallel with our increased knowledge! - And this is the current "last word" of our Universe:

It turns out that the Universe is really enormous! Much, much bigger than suggested! It has no beginning, or at least not the beginning which the current "scientific theories" suggest!

There will be no cold dark end of The Universe, which the inflation theory suggests because the Universe is constant and is here to stay! The Universe is constructed with vision, logic, precision and self-balancing mechanisms where the matter-energy exchange provides its rejuvenation and dynamic eternal existence of its structures!

I believe that this result will give the answer and satisfaction to all intelligent and honest people, which will be happy to know the truth in what kind of World we are really living and what the future of the Universe is!

The ultimate secret of the universe:

The Universe is a very sophisticated system based on logic, harmony, and unchangeable principles. Philosophical approach and consideration of the known facts is a much more appropriate method than mathematics. The mathematicians, which currently dominating astrophysics, has no chance to find and explain how The Universe is constructed and how works because mathematics and physics are based on very different and incompatible principles. To understand the Universe, we have to look in the grand scale and make logical consideration of everything we know! - To be able to get the correct picture, we have to consider the facts, and all the facts without any preferences! I would like to explain that the carefully chosen principles of the universe are the fundamental idea behind everything! We have to understand, that on the front of us is the best example of a system, which provided stability and harmony on such complex structures like the universe to exist in dynamic balance and stability for billions of years.

This system is based on three simple unchangeable fundamental principles: - Balance, Stability, and Ethics.

This is an ingenious formula where everything is working together without confrontation or a clash of interests. And if this system can be instrumental in regulating such an incredibly complex structure like The Universe, definitely, this system can be implemented as our fundamental principle as an organizing system of our society! We, the humans, have tried every possible way and many social systems, but never succeed to have peace, stability, and prosperity! We have to understand the simple truth that the attempt to cheat the others is an act of cheating ourselves out of the most valuable principle which intelligent beings can possess! – The Ethics!

Ethics is the first necessary principle, which makes it possible to achieve the other two elements – balance and stability! Without ethics, nothing works! It is time to realize that ethics are a fundamental principle of the Universe! An intelligent consciousness mind is the apex of the Universe, and ethics is the apex of the intelligent consciousness mind!

We have the best example in front of us - how those three simple principles are providing such harmony and balance to the incredible complex universal structure.

This system is proven to work and will work for us too, but I am afraid that with every passing day is getting late and late for us to change the direction where we are going, and very soon we will cross the point of no return! - Why?

What is stopping us from taking the right direction? What are we protecting? - The interests of the crooks or the future of our kids and humanity?

In the previous chapters, we briefly touched the structure of the Universe, and the logical conclusion was that quantum information and the law of physics must have pressed the existence of ordinary matter and they are the actual cause for the creation and existence of matter. The fact that the quantum information pressed the matter and is unaffected by any physical conditions as heat, pressure, gravity, electromagnetism and time leading to the logical conclusion that the quantum information is separated entity of matter and is not product and is not the property of matter! It can be the product only of the universal intelligent consciousness! That's why the quantum information is indestructible and unaffected by any physical conditions! And the universal consciousness is the actual origin and storage for the quantum information! The particles are not possessing an informational processor or informational storage capacity to remember their spin when they are in the form of waves! This telling us that:

The Universal Consciousness is storing the quantum information and is controlling all physical process in The Universe! - **This is the ultimate secret of our Universe!**

This is the secret, which the mathematicians cannot and don't want to understand and it is the knowledge, which the elite don't want us to have it!

THE SECRET OF THE ANCIENT HEXAGRAM - THE MESSAGE FROM THE PAST... OR FROM THE FUTURE?

I will start my explanation with one interesting question: What will be the simplest message if we, humans, would like to pass a message to another intelligent society of the Universe? - Just message, that we exist and we are intelligent?

To send pictures, books, or written explanations will be meaningless to another civilization because they could have a very different way of information, and probably they will not be able to decode anything, which is written or need special devices the information we are sending to be reed. It is obvious that we have to send them something very simple, common, basic, and well known to every intelligent civilization. The most appropriate form will be to be in the form of simple graphics, without text, numbers, or pictures. For example, even the two images below could be difficult to be understood by another intelligent society.

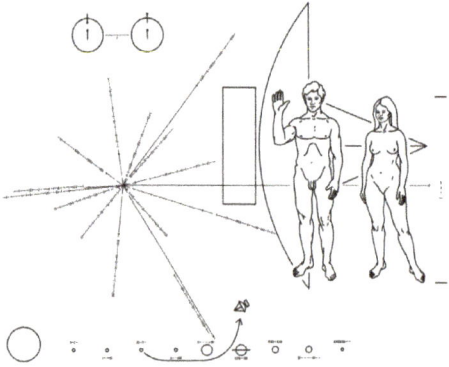

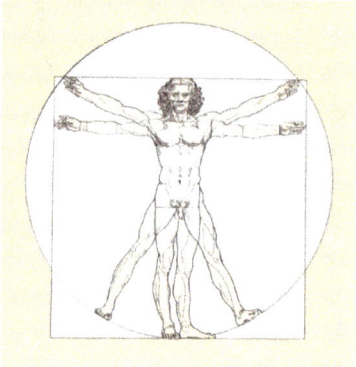

I will give you an example, how messy is our first interstellar message - The graphic on the left is the "Pioneer Plague" send in space and representing our position in Milky Way using the allegedly "known" distance to some Pulsars. The problem is there, where we have found that the light redshift measurement is not representing the rate of retreating and distance. Edwin Hubble, who become famous and is instrumental for this method of measuring distance, wrote a letter to US Astronomical Society to denounce it, but this time he was ignored. The famous astronomer Halton Arp has reported, that Quasar in close proximity to the galaxy and physically attached to it, according to the light redshift measurement is: that these two attached to each other objects have to be two billion light-years distance of each other, which is not the case! For reporting, this has cost him his job and deny access to any US telescope. He was banned from teaching astronomy and having a job in the US. So... it is obvious that we are giving a totally incorrect position of our Solar system by using pulsars! The second problem with Pulsars is that their pulsating beam is visible only in a very narrow section of the sky - they are visible only if the observer is in the exact path of the narrow beam.

Image of Pulsar and its narrow beam

Next problem: the two small circles on top of the plague suppose to represent the Hydrogen atom state of transmission to indicate the length of the time we are using. It's OK, but where is the indication, or explanation, that these circles are the Hydrogen atoms? With circles, you can represent a thousand different things! How really the Hydrogen atom should be represented, you can see on the image two pages below.

This way of drawing is a clear example of unnecessary complicated, incorrect, and difficult or impossible to be interpreted message! It shows clearly that we are not intelligent enough to create even a simple drawing, to give our correct position and to explain who we are. Or in short them, we declare that we have some technical capability, but we are obviously dumb!

Such a message should be just a simple graphical representation of something well known to everybody.

It has to be correct, simple, and fundamental. Your ability to send, and construct such a message will revile your knowledge, your intelligence, and your capability! The fact that you are sending physical message indicate, that your society is still on the primitive stage of development, and has not yet mastered the instant quantum communication.

There will be no more appropriate symbol of the basic (or the fundamental) diagram representing the fundamental elements of the Universe. - Such graphics will be well known and will be easy to be understood from every advanced intelligent society.

Our current problem is there, where we still don't know how to produce, how to recognize, or how to read such graphical representation of Universe because our understanding of the fundamental elements of the Universe is insufficient and are totally wrong! The truth for understanding of our Universe is hidden behind the curtain of the established scientific dogma, which forces us to accept the weird medieval concept of Big Bang, and the nonsense of the two officially recognized theories. - The Theory of Relativity and Quantum Mechanics, which is conflicting with each other and is producing absurd results and values as infinities, mass-less particles, and non-existing "invisible" cosmological substances. Currently, this nonsense is very well funded and is promoted as "triumph of our intelligence"! - This is the reason why we still not have a basic understanding of the fundamental elements of our World.

Without fundamental knowledge, we cannot produce or understand even the simple formula, or a simple drawing of the Universe even if extraterrestrial intelligence is sending us such drawing.

I am happy to inform you, that the time of promoted nonsense is ending, and we finally have the scientific theory, which is explaining these vital aspects of the fundamental elements, forces, and interactions of our Universe. This is the so-called - "Theory of Everything" produced in 2017 by the author of this book. The theory is included in the later chapter - 'The ultimate secret of the Universe.'

The fatal flaw of the current view that Space is Three Dimensional is laying there, where all of the mainstream "scientists" are not realizing, that they are putting the reference point outside of the considered volume of space - (in the corner of the box). The problem is there that in space is not corners, but we are inside - in the middle of the Space, and the minimum number of dimensions capable of forming and defining each point of space is Six Dimensions! - East, West, North, South, Up, and Down. On the diagram below is visible, that three dimensions can produce only 1/8 of the total volume of Space.

From the diagram below, getting clear where all the matter in Universe coming from! - It is a result of the cancellation of the opposing directional momentum of the space dimensions.

The new understanding is that Space is formed of Six Single Dimensions, which is representing the Six well-known World directions: East, West, North, South, Up, and Down. This new understanding of the fundamental elements of space is explaining, that Space is formed by two sets of (3) opposite directionally

oriented dimensions, because of East, North and Up having opposite directions to the other three dimensions: West, South, and Down.
(See the graph on the right)

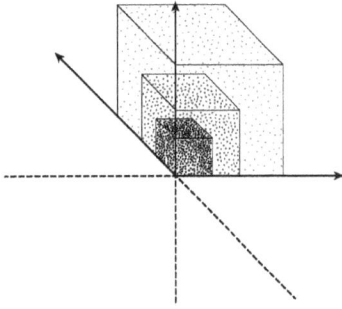

| Three dimensions can define only 1/8 of the volume of space | The graphical representation of space |

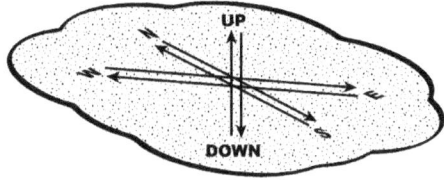

The single space dimensions are incorporated with each other and are producing the uniform space of our Universe. The understanding that single space dimension is just a vector giving us the knowledge, that three incorporate with each other space dimensions can be represented as a triangle. So... space can be represented as two triangles incorporated with each other and having opposite direction! (See below)

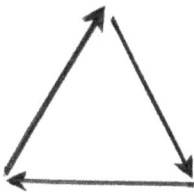

Three incorporated dimensions

The two sets of (3) opposing each other dimensions where each side of the triangles is representing a vector of a single dimension

When we are talking for "the fabric of space," we have the idea that the space dimensions are incorporated as woven fabric.

Newton's fixed space   **Einstein's flexible space-time**

The two images above are a simple representation of the "fabric of space." Exactly this is represented in the structure of the Hexagram, where the sides of the two triangles are "woven" in each other. The direction of triangles is representing the universal symbol of direction -
(Up and Down). - See the diagrams below.

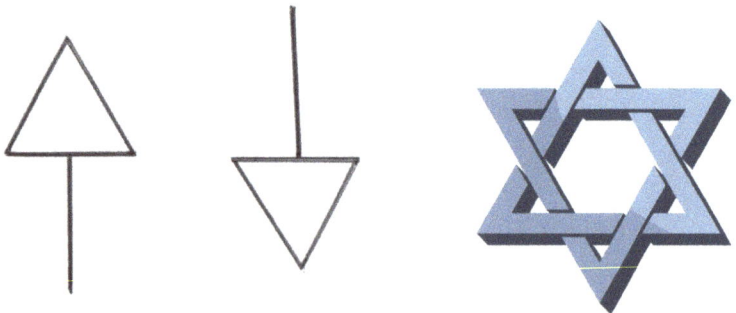

The two triangles (set of 3) opposing each other space directions are incorporated exactly as woven fabric and are perfect an unmistakable symbol of our six-dimensional space!

The structural form of matter provided by the "Theory of Everything" is explaining that the "planetary model" is the base of all material structures of the Universe. The elementary particles, when existing in "solid" form, can be represented with center bulb, surrounded by a circle of the standing waves. We can see this universal pattern in the microscopic world, planetary system, and galactic formations. It is obvious that this is the common basic graphical representation of Matter! - See the images below - section "A."

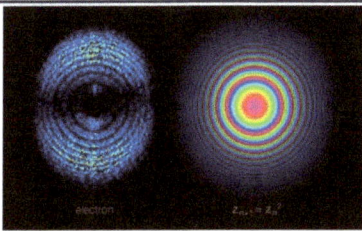

Photo of electron

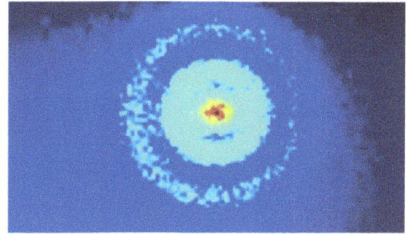

Photo of the Hydrogen atom

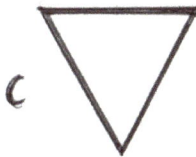

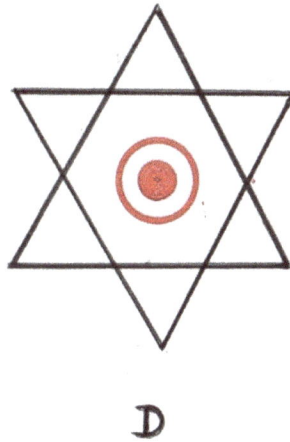

A

B

C

D

We already have discussed what is the simple graphical representation of space dimensions and matter. They are the images "A," "B," and "C." Those images are the simplest graphical representations of the fundamental elements of our Universe - Space, and Matter. This graphical representation of the Universe will be clear and easy to be understood from any intelligent society, having a correct understanding of the physical property of the

Universe.

And surprisingly, when the simple graphical representation of the Universe is complete, (Diagram "D"), we had on the front of us the most common and widespread Religious Symbol on Earth! - The Hexagram!

The image of Hexagram is originated and found everywhere in the deepest layers of human history and many different religions. - The first images of the Hexagram we have from about 10,000 years BC.

Batkona star   Hexagram in ancient Mosque

The carved in stone images of Hexagram we found in Lebanon, 10,000BC, Egypt, 10,000BC, Sumerian tablets, Baalbek temple, Cyprus, 5,000BC, Mayan stone carving, in Kurdistan, Tibetan coins, Mongolian coins, Moroccan coins, and is all around in Catholic and Orthodox Churches and Cathedrals. The symbol of Hexagram is a very important symbol for Vedic religion - (Shatkona) Hindu, and Buddhism

| Hexagram in Kathmandu - Nepal | Hexagram in Japanese scroll |

The Hexagram is regarded by the ancient religion <u>as a cosmological diagram or symbol of creation</u>. The appearance of the same symbol all around the World, representing the creation of the Universe cannot be co-incident. To be regarded as one of the most important symbols of all religions, this symbol must have its real and common origin! And the origin we can find on many ancient texts and stone carvings. Surprisingly, the ancient texts insist, that the hexagram is given by the "Gods" which has descended from the sky and has giving humanity vital knowledge. (I am not going in mysticism, I just follow the archaeological facts)

(see the images below)

Sumerian stone carvings, revealing the heavenly origin of the hexagram

In the present time, the Hexagram is promoted and commonly is known as the "Star of David," but David has nothing to do with this symbol! The Jewish king Solomon has using this symbol as seal only. The Jewish start using the Hexagram as an official symbol very late in time - since 1648 in Prague. Later the Rothschild has adapted it, and finally, the Zionists adapted the hexagram as their official symbol in 1899. Currently, it appears on the Jewish flag. - A vital clue for the value of this symbol coming from the fact, that the Jews are commonly attracted to the items with value. Even the shops selling gold are named as (JEWelry) shops.

They are in possession of the finances of the World, controlling the Share Market, the Gold and Diamond trade, and is taking as their own - the most valuable symbol of the World - the Hexagram! It will be good to know that the origin of Hexagram is coming from the deepest history and has nothing to do with the Jews! They are free to adapt and used this symbol, but they have to know its real origin and meaning and not to claim that it is just star of David with an empty center! - The Hexagram without a center is as the body without soul and heart. We must keep and preserve its original form and meaning which contain seven elements - six space dimensions and the matter - it is the symbol of the Universe! - Universe filled with life, meaning, and purpose, - not just an empty Universe! It is important to know that Hexagram belongs to all humanity! - It is given to us with special reason! We have to realize that the Hexagram is an unmistakable and simple geometrical diagram of the Universe It can be created only of intelligence with advanced knowledge of the Universe! Even in the present time we still not possessing such knowledge. Thanks to the new theory, we are just opening a window of knowledge to find the real wonders of the Universe!

Such enormous blocks cannot be lifted by manpower and wooden sticks

Such complex shape and precision cannot be achieved without laser cutting and computer

Granite cannot be cut without hardened steel tools and diamonds

The ancient advanced megalithic structures, including enormous perfectly cut granite blocks, which cannot be cut with bronze tools. The complexity of forms and precision cutting of these megalithic structures cannot be achieved without diamond tip tools and computers. We didn't find any steel tools, and the logic is pointing again to the fact that extraterrestrial visitors cannot carry stone cutting machines, but they will have possessed lasers and will be able to use antigravity to move these enormous blocks on big distances. The sudden appearance of agriculture, irrigations, and advanced celestial knowledge of the invisible planets of the Solar system cannot be explained any other way than this knowledge has been given to us.

On the carving are visible the known nine planets of Solar system. The celestial calendar is too sophisticated to be invented without any prior knowledge.

This obviously is given knowledge, because to know this, you need to have powerful telescopes and precision measuring instruments.

Such complex parallel advances in knowledge, capability, and social organization cannot be the achievement of primitive hunter-gathers. If the logic is pointing in this direction, we don't have to be afraid to recognize, that all these sudden explosions of craft, technology, knowledge, and organized societies, in combination with the appearance of the secret religious symbol of Hexagram, can be explained only with interference, and temporary presence of extraterrestrial visitors. The Hexagram has been left for us - (the future generations) from the visitors of another advanced civilization as confirmation, that they have been here in recent history and has been all around the World. They teach the early humans how to do agriculture, how to carve and cut stone, how to make irrigations, and how to live in peace and harmony. The visitors from another star has been found the emerging human civilization in a very primitive stage, and they have done as much as possible to kick start one more ethical intelligent and humane society. They have given us as much knowledge we have been able to absorb, understand, and use. Knowing, that "They" cannot stay for longer, and probably they will never be able to return, ET has left a message to us - a simple message, which we will be able to understand as soonest we obtain the basic understanding of the World. In the time, when we had no writings, no numbers or any understanding of the World, "They" has no any other way than to live here a message for the future generations - message as a simple drawing of Universe with specific instruction to the ancient people - this symbol to be preserved as something very, very important! They knew, and hope, that one day we will become much smarter and knowledgeable and will start asking the question: Are we alone in the Universe? And if the answer is that we are not alone, where the others are? Why is the Universe silent? - This "simple" symbol has deep meaning and purpose and will explain to us a lode! It is our lifeline! This simple drawing is giving us the knowledge that we are not alone in the Universe! This drawing is telling us that space travelling is not a myth, but it is a reality! With this symbol, "They" have left a simple message to us: "We are intelligent. We have been here. When you are ready, just contact us!" They have left this message with the hope that when we realize, that we are not alone, we will start asking ourselves: why, and what is the reason the other intelligent societies not to revile themselves and is not responding to our messages? "They" was hoping, that we will become smart enough and will be able to understand the simple logic and the real meaning of the message: That they have been here and was teaching us how to live in peace and harmony. They are remaining us again that cruel, corrupt and destructive civilizations are not welcome in the United Intelligent Society of the Universe. Even the first interstellar object - "Oumuamua" which recently passed through the solar system has all features and behavior to be space ship - elongated shape, not emitting or absorbing anything, and on the way out it change direction and accelerate. (It was silent too!)

"They" has been hoping that we will be able to realize, that the Universe is based on unbreakable principles and ethics, and if we would like to continue our journey, we have to start to obey those principles.

It is obvious that we still are not ready to be contacted, but the clock of our destiny is ticking! Can we be able to qualify before our clock stop ticking?

The symbol of the Universe and the stars around

I would like to consider the nature and potential of two major threats for our existence – quantum computers and bio-engineering:

Quantum computers

What are they? And why should they threaten our existence?

The quantum computer is based on absolutely new physical principles. To be understood better, I will start with the basic constituencies of matter and their functions:

Quantum mechanics in the early 20thcentury found an amazing phenomenon – that particles of matter and the free particles in space are connected by a mysterious information link. This information link is not affected by time, gravity, temperature, or pressure. In short, quantum mechanics states that the most stable property of matter is its information. This information cannot be created, cannot be destroyed, cannot be copied, and cannot be duplicated. I have discussed all these conditions and properties of the quantum, or (universal, information). It is one of the major clues for an intelligent origin of the universe, but we will leave this subject aside and will consider only the practical aspect – how this information can be used to create a very powerful new type of computer.

There is a phenomenon in particle physics called quantum entanglement. This phenomenon is the mysterious information connection of pairs of free particles. The information link between them travels with instant speed, no matter how far apart those two particles are. Even if they are on the opposite sides of the Universe, the information knowledge of each other's property travels in an instant. Quantum information can travel backward and forward in time when it needs to adjust and balance the property of matter. (We have discussed this in a previous chapter) - That the position and velocity of the connected particles in space put them in a different time, according to the Theory of Relativity, but the information link between them is not affected by their local time difference.

It is exactly this phenomenon of information-connected particles the scientists now want to use to produce a very powerful new type of computer. Currently, scientists face technical problems with how to contain these particles. The necessity of super-low temperatures and the absence of energy and radio emission is a big obstacle for them. In the Earth's environment, where we are emitting all kinds of radio noise and other kinds of physical interference, from which the experimental computers have to be shielded.

I would like to turn your attention to these necessary conditions in which the quantum computer needs to be able to work – low temperature and absence of interference because this is the exact conditions of deep space! In these perfect conditions of deep space, even if the quantum information originally hasn't been organized, <u>in these conditions, the existing quantum information in the universe will inevitably organize itself into a more intelligent form of a super-intelligent quantum computer, - (or as universal intelligence</u>)! - (This is a crucial fact of our knowledge which is carefully hidden from us - the public). <u>Even if for some reason the universe had a spontaneous origin, the information of the universe had all the necessary time and conditions to organize itself as super intelligence of universal scale and to exist already as super intelligence long, long time before human appearance!</u>

The significance of this revelation is enormous, and I would like to leave it for the readers to make their own conclusion.

Let us continue with our subject of a quantum computer. The basic idea is to use the properties of those connected particles to process and store information. In a classical computer, the basic unit for functions can be 0 or 1, but the amazing properties of the connected particles are able to be simultaneously in the two states, 0 *and* 1. This increases significantly the potential computing power of the new computer and increases progressively with the number of particles involved in computing. Currently, the experiments involve only a few pairs of particles, but if the technology overcomes the technical obstacles and is able to involve, say, about 500 pairs, the ability of this computer will exceed the extent of our best imagination. Puzzles and calculations, which our current most powerful computers need billions of years to compute, will be solved in minutes by this computer with its problem-solving ability, logical solutions and simultaneously parallel calculation of all possibilities! All this enormous potential of this new type of computer will allow it to work out how to upgrade itself with an ever-increasing rate and become independent from us! Computers with emerging intelligence are a very dangerous scenario. <u>Lower intelligence will never be able to control higher intelligence.</u>- This is an undeniable fact! Such intelligence, far more superior than our intelligence, with self-upgrading capability and sophistication, and without the necessity of our intervention or help is called by some scientists - 'our last invention,' because we will have created something superior and more capable than us whom we cannot control! This intelligence will not necessarily be grateful or loyal to us because machines have no emotions, ethics, and loyalty – just logic! The logic which we apply to harmful parasites - (to exterminate them and get rid of them), is the logic that will apply to us, because we are exactly parasitic destructive creatures, and we are an unnecessary element for nature and beauty of planet Earth. Nothing will stop this artificial intelligence to continue upgrading itself and obtain the capability of universal scale, and there is the next potential problem:

If there is super intelligence or creator of the world do you think that "He" will allow us to create intelligence equal like "His"? (I don't believe "He" or anybody in "His" place will allow this to happen).

Here I would like to explain the logic and vision of the creator (or the superintelligence) and the measures implemented to prevent this to happen: Above we have considered the point where AI will obtain capability and computing power on a universal scale. But there is the next embedded limit which will prevent AI to become new "God" The key and the limit coming from consciousness. - No matter how sophisticated AI will be, AI will always remain an unconscious machine! We cannot give consciousness to a machine! Intelligence and consciousness is a two fundamentally different thing! There is a carefully chosen limit of knowledge for AI. AI will be able to obtain everything we know, but this is not enough because our knowledge is restricted by the "limit of knowledge." The limit of knowledge for AI is coming from consciousness by self! - The consciousness is the actual storage bank for the universal quantum information and the law of physics! Without consciousness AI effectively will be cut off from the storage bank of Universe and from vital information for the structure and property of mater and Universe! And exactly this will prevent AI from becoming the next creator or destroyer of the world! AI will be able to deal with us, but won't be able to mess with the order of Universe.- This limit is completing the chain of logical limitations for destructive knowledge and capability! We have to realize that nature is not putting such logical conditions and limitations, but the intelligent mind does!

Bio-engineering:
Scientific advances in the knowledge of DNA, cell functions, and structure give us a new understanding and the ability to interfere with the functions and structure of living organisms. The complexity and functions of DNA are enormous. The way this information is stored can be compared with a chain o all the books in the world arranged in a strict continuous order. The combinations of this information order are beyond any current possibility of being calculated or understood. Currently, scientists are able to find the specific sections of DNA responsible for the individual functions of organisms. By experiments that alter a part of DNA and use directional mutation of selected bacteria, military laboratories are able to produce many new types of lethal sicknesses with the potential to spread fast, to be resistant to antibiotics, and kill indiscriminately on a mass scale. Despite the international ban on biological weapons, the developed nations continuously are developing and stockpiling new and more lethal biological weapons. It is only a matter of time when some accident will happen, or some bad or mad person will open Pandora's Box. The results and consequences for humanity are unknown. Some people leak information that the elite are seriously thinking of and preparing for Earth's depopulation. If we don't want such a scenario to happen, we should effectively and unexceptionally ban such shameless activity, because the reason for developing such lethal bacteria's is to obtain the capability of indiscriminate killing on a mass scale! Unfortunately, we are not in a position to stop that until we are not in effective control of our

governments and our society.

The other aspect of bio-engineering is the work toward creating methods for intervention in the human body systems. The potential for new ways of treatment of many sicknesses is enormous! But the coin always has two sides! Treatment is one thing, but genetic modification of unborn babies and altering future functions of organs and body functions and modifications are absolutely different things! The fact that we need probably thousands of years of careful study and experiments to be able just to understand the mechanisms and functions of DNA tells us that it is not wise to play with something we do not understand. The consequences of this can be tragic!

More than obvious is that the elite are using our tax money to fund research in the area of genetically extending human life. In the DNA of each body, cell is imprinted a code for the renewal of this cell. When those programmed renewals are exhausted, the cell gets old and eventually dies. If this pre-programmed obstacle is removed, in practice, this cell will be able continuously to renew itself and will become immortal. Scientists already have found and know the chromosomes responsible for this mechanism.

The fact that the genetically modified seeds of US - Monsanto cannot reproduce and every year farmers are forced to buy new seeds is proof of the existing current knowledge in this area. The breakthrough in extending human life could happen any moment, but who will benefit from this and what will be the implications and consequences for all of us? Sound exciting, isn't it?

But reality and wishes are two different things! If such an invention happens, the first who will benefit and extend their life will be the super-rich. There immediately will be inserted a high price barrier, which the ordinary people cannot afford. The next thing will be that this invention will also be used as a bribe for politicians, army generals, police chiefs, or people in key positions. This will give elite unlimited power and absolute control. Such situations will establish a terrible regime of indiscriminate tyranny and cruelty because we already know the "ethical" and "moral" standards of the elite. When the extended life of the chosen minority effectively starts creating overpopulation on Earth, then by stealth and silence they will start releasing on a regular basis the biological weapons for Earth's depopulation. Don't have any illusions! 'They' have to do it! 'They' will have no any other option, because there will be not enough room and resources on Earth! - This is our tragic reality, and we are getting there very fast. – **The future of uncontrolled scientific advances in the hands of corrupted and unethical minority is a grim prospect for the future of humanity.** "They" have a plan! - A good and hidden plan, but for the rich people only! And there is nothing good in this plan for us, - the ordinary people of planet Earth! They have pass legislation for compulsory vaccinations of our children. The claim, that the not vaccinated kids will spread epidemics is not credible, first, because the vaccinated kids don't suppose to get sick, and the second fact is that humanity manages to survive, without vaccinations, but with good hygiene and isolation, we can prevent epidemics. The reason behind the compulsory vaccination of kids is "They" to have legal rights to inject biological and DNA material in our kids, without our ability to oppose, check what they are injecting, and stop this activity.

Our real tragedy is coming from our reluctant acceptance of the corrupt system and our intellectuals, who have betrayed us and have joined the elite just to have a job and financial security. In this crucial moment for humanity, instead of leading us and protecting us from the emerging social and technological threats, the leading scientists reluctantly have joined the campaign of misinformation and fake science!

We still have a chance, a very little chance, but still, this is some hope - to get together, organize ourselves and take the power from the elite in our hands before the catastrophe happens. The only solution will be to create and implement uncompromised ethical standards of a democratic, ethical society where the control over scientific advances will be in public hands, and the benefits of them will be for all of the society, not only for a selected minority. We have only a miserable number of years before it becomes too late. If we don't take control now, our future is far from certain! – And then, when our signals stop, the Universe will become even more silent than now! - Why?

BIRDS NEST, ETERNITY AND REALITY

The dream for eternal life, immortality, and endless happiness are as old as our civilization. Many legends, predictions, and attempts have been made, but this dream still remains elusive. Probably we are much closer to achieve this dream than anybody really realizes. Genetic engineering is well funded and at any time could deliver a breakthrough in extending life by making our aging cells able to renew themselves continuously into young ones. By programming ourselves this way, life expectancy could be limitless. Our technological advances can provide us with everything we need – (if we are not greedy and stop robbing each other).

But we have one problem, one big problem! I call this problem the 'birds nest' problem. What do I mean? It is simple! If growing chicks do not leave the nest, there is no room for the next generation.

In the same way, if we extend our lifespan to, say, one thousand years, there will live together 16 generations on Earth, a total of 116 billion people, and the number will keep growing. It is obvious that the resources, our technology, and Earth cannot sustain that big population. But there is 'good' news. We don't have to worry too much, because our super-rich elite has already planned for this! They will limit life-extending technology to people with enough money only – (themselves). At the same time, they are silently preparing plans for Earth's depopulation in the form of wars, plagues, genetically designed deadly epidemics and the forceful or secret genetic manipulation of ordinary people's health and life expectancy. When there was an accident with the Ebola virus, the Pentagon was forced to admit that they had supplied about 300 laboratories in the US with the virus. This is only the tip of the iceberg; this is the only one of the thousands of dangerous bacteria that the military is preparing for us! The over-population of Earth will be not an issue for them. The elite relies on the total control of media and misinformation. They rely on people's reluctance, obedience, and the widespread belief that we are living in democratic countries and there is no better system than ours and nothing can be done to change this tragic situation, where people have only rights to vote for one or another corrupt politician, who serve the elite interest only!

I believe that when we were created, our creator had different plans for Earth's nature and for us and our future. Our greed and acceptance of corruption is our main problem! We are going in the wrong direction – (same as science), and there is no way out! Our creator desperately is sending us messages for help, but we are not listening! We are given the freedom to choose to do nothing and accept the elite to be our new 'gods'–'gods' proven to be greedy, ruthless, corrupt, and cruel. However, we could choose to organize ourselves and take our destiny in our hands and have a future. Then we could choose another scenario for solving the bird's nest problem and the future of Earth and humanity.

There is a way to achieve the dream for long and happy life. The first step is to adopt ethical standards and establish real and protected from corruption democracy, where people will have the power to determine their lives, their future, and destiny by themselves. The second step is to estimate honestly and realistically with the current technology, what is the maximum human population that can live on Earth in harmony with itself, nature and other living creatures, and not exhaust Earth's resources. Then we have to accept the necessity not to exceed that limit and the quota of 2 children per couple. Then later, when we have better technology and people have the choice to extend their life, they can choose for themselves whether to have a normal lifespan and normal life on Earth, or to extend their lifespan and, when they reach the age of 60, to leave Earth and colonize other planets, and create a new life there in space, - new societies with uncompromised moral values based on ethics, love and happiness from the very beginning!

This is not a dream; this is a reality, but a bitter reality, a cruel reality, because at the moment we have in place only the first plan, - the cruel and senseless plan of the corrupt elite! - A plan where we have no future!

PURPOSE OF LIFE AND ETERNITY

Why we exist and what is the purpose of our life? This question always has fascinated me, and I never found a good satisfactory explanation for it. When we are looking at the way how nature and the Universe are design, we cannot deny that in every detail of the world around us is embedded logic and purpose. There is no one particle of matter which is not involved in the physical order of the Universe, and there is no one bacteria, plant or organism, which is not part of the well-balanced bio-system on Earth. It is obvious that we are sitting on the apex of biological life, and every one of us is part of this grand design. Everything, no matter how small and insignificant it is part of some system and has a role to play. It is difficult to find what really is our personal role and purpose in this amazing system, which we are calling World. The simple concept of the religion that our purpose of existence is our happiness has never satisfied me and never make logical sense. This explanation actually is not giving us any purpose and role but is disconnecting us from the dynamics of the world, putting us in a cocoon of insulation and selfishness, it is diverting our attention and efforts toward us, it is disconnecting out mind from the real world! This doctrine doesn't make sense, because the Universe is very logically built and to be the only highly intelligent creatures on Earth and the assumption that our purpose is to serve only our self-satisfaction doesn't make any sense to me.

The idea for our real purpose in life starts crystallizing in my mind when I was watching a presentation of the famous biologist Bruce Lipton. He is explaining that in the development of living organisms is embedded logic and continuation of sophistication. His examples follow the developments of life on Earth. There is an obvious trend of continuous sophistication in life forms development - from the appearance of the first primitive bacteria, followed by more complex and sophisticated single cells possessing DNA. The next stage of sophistication is the formation of bacterial colonies. Those colonies are collectively organized bacteria, which share the benefits of collective existence, share food, nutrients, information, and collective defense.

The next stage of sophistication is the appearance of complex cell organisms – (us) we are an assembly of highly organized single cells, where each of them has specific roles and duties. The single cells of our body are enjoying the great benefit of the collective order of our body – they have constant food supply, constant environment, protection, care (if they get sick), and much longer life expectancy. The collective organization of our body cells is giving them enormous benefit compare to the single existed cells, but the most important things are the new level of capability of the complex cell organisms. We are incredibly mobile, smart, and able to sense and evaluate the environment and to repel any threat to our existence. The capability of complex cell organisms compare to single-cell organisms is incomparable. - The jump of sophistication is amazing! Bruce is also pointing, that our existence is not the last word of sophistication and that there is the next level of biological development, which will increase our capability again on a new higher level. The next logical stage again is the gathering of the individuals in a highly sophisticated new form of super-organism as an organized conscious linked society, where each member will enjoy the great personal benefit of the collective order, which will give each member great access to the wonders of the Universe. Bruce logical consideration is pointing toward the next stage in sophistication, where the consciousness will provide the necessary unity and great collective power and capability.

We have already good examples in nature – The collective order of bees. Each bee has its role and duty in the collective assembly, but the benefits for each of them are much greater than their personal contribution. We have to learn from these small amazing creatures how they live in peace and treat each other with tenderness and love because their example clearly shows that if we treat each other with love and respect, we can achieve much more, than when we are fighting and robbing each other. Bruce Lipton logical consideration is pointing that the next stage in biological sophistication inevitably is exactly this – forming consciousness united society as one super-organism, with enormous capability and sophistications, where the consciousness will provide the necessary bond and unity and will give us new level of mental power and capability. More or less the consciousness is the creative power of the Universe and unification of our collective consciousness will give us enormous creative capability and understanding of the Universe.

This understanding of the direction where the development of biological life on Earth progressing had helped me to understand and formulate the answer to the vital question of our own purpose of existence - or the purpose of our life.

One day, when I was reading a book for the evolution of life, the answer to this question suddenly pops out and crystallizes in my mind. I was stunned; it was a fantastic realization of an idea, which makes absolute sense and explains, and giving meaning and purpose to the existence of every one of us. Will be better if I start the explanation from the beginning:

As usual, the institutional church, philosophers and pseudo-science giving us the senseless and incorrect answer to every important question of our life. This also is the case with the question of our purpose. - They give us their egoistic, selfish agenda that our main goal is to try to be happy at any cost, despite the pain and suffering their own satisfaction is causing to our society, and where we must be passionate and to forgive them everything they do to us!

Suddenly, with disbelieve, I have realized that behind the very nice and well-formulated slogans that the purpose of our life is our happiness is hidden the root of our suffering and misery! –they are proposing happiness true, sacrificing the ethical and moral principles in combination with blind obedience! –Principles, where if somebody smacks you, you must turn your face and give him the other side to smack you again! - Is this make sense to you? Is this being the way how we should treat injustice? Is such inaction will give us happiness and purpose in life? - Absolutely not!

If you are not going to analyze carefully, this philosophy looks like very nice, good ethical behavior which supposed to bring love and peace in our society, but what really this behavior brings us? - Endless suffering, injustice, wars, and continuous eliminating of the most bright, valuable, and intelligent people of our society. This way of "ethical" living is nicely formulated, but have deeply hidden agenda, which serves only the purpose of the rulers and bringing them wealth and unopposed power. This exactly is what the corrupted like – everybody to be selfish and to try to please himself only, to be calm and obedient, and to be tolerant of cruelty and injustice. I nave realize, that this philosophy is the actual instrument, which is striping us of basic understanding of what our role and purpose of life really are.

When I have considering the trend of biological evolution, I have realized, that I am a link in the chain of our ancestors and I am the connection to the future line of our family generation. I have realized that all my ancestors have been successful for many millions of years and have succeeded to survive any cataclysm, any disaster, wars, famine, and succeed to give me life. At this moment, I realized that my life is something unique and special, and I don't have to waste it because I also have duties to future generations. I have realized that my duty is to provide a successful link and continuation of our family line and to pass the knowledge and capability of my ancestors to future generations. This realization was enormous because suddenly, I found the logic and purpose of my existence and my active role in the order of the World.

If every one of us is living just for himself and not making any impact and contribution to the world, our existence will be useless, will be no any difference if we haven't existed at all.

The pharaohs have to try to achieve eternity, by hiding their dead bodies and possession in the pyramids – they didn't succeed! – They got robed, and even the bodies of some of them become museum exponents for fun of the public. Many have tried to copy them, but mummification of your body cannot give you eternity! The way of eternal existence already is provided to us! - The biologists say that our body is mortal, but the information, which DNA carries and is passed from generation to generation is immortal! And this is the way for us how to ensure our eternity – by rising our future generation, care for them, teach them our values, our ideas, giving them our conscious spirit of the universal ethical principles of freedom, love, equality, justice and passionate! **We have to learn, preserve, and pass to future generations the knowledge and the ideas of our ancestors, and add there our own unique personal contribution to those values! We have to teach our kids and give them ideas on how to live and how to make this world a better place! - This is the purpose of our life and our conscious, intelligent existence! - To live behind our own mark, our spirit, spirit possessing all the quality of free, intelligent and ethical conscious human being!** -This is the way how to preserve and pass our spirit to the future and achieve our eternity!

This is what makes sense to our existence, and have nothing to do with the selfish, senseless greedy philosophy of the corrupted elite.

But don't understand me wrong, I am not opponent of personal happiness, it is exactly the opposite, I believe that everybody deserves to enjoy the life and to be happy, but this doesn't have to become an obsession and sole purpose in life for the expense of the values and ethical principles. Happiness has to be the end product of our efforts, not the senseless happiness in the form, which currently is promoted, but happiness to be an active, ethical and constructive element in the dynamic order of our World. We have to create a simpler way of life, where every person will have the motivation to learn, to work, to create, to love, to care for their kids and parents and to be happy with his achievements! This is our biggest treasure and the meaning of life which we have to achieve! Not money and power to suppress, rule, and exploit the others! We have been given the most beautiful planet in the Universe, teeming with life and unimaginable beauty. We have been given the treasure of intelligence and the internal world of feelings, emotions, love, and ambitions. We have been given everything that we need to live in peace, happiness, and harmony! And we have to be wise to preserve, share, and enjoy those gifts that we have been given!

In order to survive the ultimate test of ethics, we have to start understanding the grand reason and purpose of human existence - to grow up as moral, ethical, intelligent creatures and form consciousness united society, to live in peace and harmony with nature and join the advanced intelligent universal society. We have to adapt ethical approach and respect to all life forms on Earth and in the Universe! The oil and coal have the same chemical composition as our cells because they are a natural product of fossilised living organisms! Instead of burning them, we can use them to produce our food. Our bio-engineering already has the ability to create and grow our food and proteins in laboratory condition and to eliminate the necessity to kill the animals for food! Instead of killing, we have become life guardians. We don't have to colonize the galaxies and saturate the Universe with ourselves only! We have to do what the others have done for us: - To spread the sits of life on suitable planets and distance ourselves, let the new life develop their unique forms and quality, which one day could bring new and rich cultural, ethical values and ideas and contribute them to the intelligent societies of the Universe! This is not a dream or utopia; this is a necessity; this is the only possible way our society to progress to the future!

Our current effort has to be directed to become an ethical society in order to survive and be given rights and freedom to communicate and contact with other ethical societies, rights to explore the Universe and become part of its intelligent order!

I am sorry, my friend, but I have to be honest and tell you the truth- all the truth, not just part of the truth. There is something else, something very important, which we haven't been discussed yet. - Some people may object and will disagree with me, but this is the naked truth! – The most valuable thing in our life actually is not our happiness but is our freedom! Freedom is the most fundamental principle of life! Not physical freedom only, but also our spiritual freedom, freedom of expression, and freedom of choice! Without freedom, we cannot achieve our purpose in life, and without purpose, there is not happiness. An only a free person can be a truly happy person!

The history is telling us, that humanity has been treated and breed on the same way, how we are breeding and selecting our domestic animals: - Only the calm and tamed individuals are allowed to breed and produce the same kind of obedient offspring. During the millenniums, humanity has been methodically depleted from its most valuable intelligent and bright people. Anybody, who dare to question the authorities has been brutally eliminated. This is the pattern of each tyrant, dictator, or ruler. The elimination of unwanted people turns up as industry: - In middle ages was the Inquisition, followed by the repatriation of political and criminal prisoners in the New World, then the Communist system of Gulags. The tyrants have using the wars as a tool for the elimination of unwanted by sending them to the front line to be killed. The history remembers cases of incredible brutalities, such as were in the battle of Stalingrad, against the surrounded German army has been sending a massive number of political prisoners with no rifles in order to be killed and the Germans to exhaust their munitions. Recently Khmer Rouge has killed millions of his opponents and got away with this mass murder because it is the agenda of "forgiveness." The agenda of forgiveness always is one-directional – toward the tyrants, but the tyrants and authority never forgive their opponents or us! – Julian Assange was held in the Ecuadorian embassy for seven years just because he exposing the high level of corruption. The result of this continuous elimination of ethical and freedom-loving intellectuals have produced the core of our sick society, where the obedience and fear to preserve the miserable job or salary become obsession stronger than any ethical and moral principles. To preserve our personal comfort, we don't want to know the truth and to question the authority – why so many nations are destroyed and billions of people living in poverty. Why is hunger when our technology can produce more than we can consume?

This is the naked truth which most of us even don't want to here, and its core is called –selfishness, ignorance, and obedience!

All slaves have one common problem: - they are afraid to fight for their freedom.

We have to realize that a free person is only the person who is not afraid to fight for his freedom! - You cannot enslave a man who is ready to risk his life to defend his principles and his freedom. An only a free person can achieve his principles and have real happiness. - Everything else is just compromised.

For comfort in life we- the westerners are paying dear personal prize – we are exchanging our moral and ethical principles for money, and not any money, but for our own money, - the money which we are earning with hard work. To justify our acceptance of corruption, we have produced the most shameful slogan: -

"All politicians are corrupt, but we need them!" This is the sign of a total misunderstanding of reality. This is the last thing in which we have to believe and accept because there is a way to get reed of the corrupt politicians, and not only of them but also of their corrupt governing system! - There is a way, good and elegant way of solid and indestructible Democracy, which is design to fight and win over the corruption of any kind! It is a simple and effective system, which is explained in the last few chapters of this book. - (The way Out)

This is the knowledge, which we have to learn, embrace, and make our goal in life! This will be our mark in the history of mankind -our knowledge and principles to be passed to future generations! This will be the best purpose of our life – to build a society free of corruption, cruelty, injustice, and inequality, to teach our children to prize and value the freedom, and how to live in harmony with themselves, with nature and the Universe! This will be our way to achieve eternity, and if there is God or Creator, we will be well-positioned our consciousness to be estimated as something good and valuable for the intelligent life of Universe and to be preserved for the eternity to be set free of the limitations of our mortal bodies and be allowed to roam the World, enjoy and explore the beauty of this endless and wonderful Universe!

Some people can say that this is just a dream, but I will like to tell them: It is not worth to live if you have no dreams and is not fighting to make them a reality! -

If the bees have succeeded, why we shouldn't?

Only an ethical society can avoid apocalypse

In this chapter, I am not making a statement about God's possible existence. I only consider the logic behind the structure of the world, and what the logic will be if God exists, creates the world and why the world is so cruel.

Many religious people ask this simple question: why does God not intervene and do something to put an end to our cruelty, wars, greed, and dishonesty? I used to ask the same question: if God exists and is able to create this world, why didn't he create the world perfect, without cruelty?

The answer is simple and logical –there is no other way! - To have white, you must have black; to have a cold, you must have hot; to have good, you must have bad. Without contrasts and opposite properties for comparison, the terms lose their real meanings.

If there is a God, he has to look at us in the same way that we look at our children. We teach and prepare them to be independent of us, to be able by themselves to decide what is good or bad and to be people with free will and ability. If we are dictated to by God what to do and what not, then we would not be free. In that case, what would be the difference between us and robots? We will not possess any moral values, freedom, or self-determination if we are dictated to. We would be only puppets, and there would be no meaning of good or bad! - This is the only ethical and correct way to create a free intelligent life. And this leads to the logical conclusion that such an order must be designed by (an) ethical and intelligent mind. We are given freedom to possess moral values, we are given the choice of good and bad, and we can choose the good or bad by ourselves without intervention and fear of punishment. Only then, when we choose what is good, we can say that we are good and possess the moral values of civilized creatures. Those intelligent and ethical principles of non-intervention do not stop 'him' giving us advice and directions, but I believe that they are given in a very delicate and ethical way also! For example, if you were doing something bad, you can have another thought or feeling which is trying to stop you from doing this. The creator does not let us know his presence, but it looks like he is always next to us; we are just too busy and ignorant and do not put any effort into feeling his presence and gentle advice.

Then, the question of an afterlife remains open and uncertain! Yet it is even more important than life itself. Nobody, including the church, can know with absolute certainty the answer to this question, but knowing that space is unlimited and there could be hidden dimensions and hidden worlds mean tha if God exists, he will have enough room to keep our souls (of course, if he chooses to do this). If he would like to keep and extend the existence of our souls, he also has a choice – and it is very logical that he would choose to keep what is valuable and discard the rubbish. This is very logical and is consistent with the logic that we observe in the structure and properties of the world. The consistency of all religions regarding the afterlife is very likely that this idea has real origin from our conscious sensitivity and the gentle connection with the enormous consciousness of the universe. From the logical principles with which the world is created, it is easy to understand that behind the scenes, there is a hidden intelligence possessing high morals and ethical standards. In one sense, nobody can promise you eternal life, but if you follow the logic in the ethical standards and generosity of the creator, it will be very likely to extend also to the soul and consciousness of the ones who deserve that. Nothing is guaranteed! I am just using logic and connecting the dots. If this is correct, you effectively can destroy your chance for eternal existence with incorrect actions. It is obvious and logical that our actions and ethical standards are the real criteria for determining our personal value.

It is logical and obvious also that we cannot blame anybody else, including God, for our cruelty and the wrong things that we do, and then ask God to intervene.

We are given freedom to kill each other, we are given freedom to extinct the human race, and we are given freedom to correct ourselves too! The most crucial fact to understand the reality of our situation is that - you cannot help somebody, who is reluctant to help himself! – The doctor cannot help the patient who refuses to take pills. You cannot help gambling addicts by giving him money! In the same way, God cannot help us if we are not willing to help ourselves and take actions to stop our cruelty, ignorance, and correct our wrong social system!

Don't beg God to take away our freedom; he will not do this! We have to do it by ourselves! The time is running out, and we have to make our collective choice. We have to take our destiny in our hands and change the world into a better one, where moral and ethical values will be our standard of living. We have to choose this now because in a few years' time it really will be too late! We are given the freedom for the ultimate choice – to live or perish. To do nothing is also a form of choice, but it really is not a wise choice!

Bombing of civilians become standard!

Such things a civilized society would never do!

After examining the basic theories for understanding the world, it is absolutely obvious that they are wrong, and are deliberately created wrongly for the purpose of confusion. This information chaos is created as a cover-up operation by the super-rich so that the elite can rule unnoticed in the chaos and confusion. The reader will ask who actually is behind all this chaos, confusion, and corruption. These people always prefer to remain hidden and to rule behind the scenes. This is a very sophisticated strategy of stealth, where the unsuspecting public cannot figure out what's going on, and blame and changes the politicians, but the actual rulers remain unaffected and untouched behind the scenes. By using logic and available facts, we can easily find out who these people or groups are who are responsible for the endless wars, confrontations, recessions, exploitation, injustice, and human misery– just follow the money!

From our deepest history, we know that two institutions are definitely involved in the present hidden order – the Catholic Church and royal families. The next group is the financial owners of the national reserve banks, major banks, and the world's financial institutions. Not far behind are the oil magnates of the US and the Middle East. In the same category can be included the US military industry and some rich individuals like Bush, Clinton, Bill Gates, George Soros, and Kissinger, who are only the front men. (It is hard to provide documentation to support this. Some, who have tried to do this, is exiled in Russia, another is in the Ecuadorian embassy in London, and the third one has

208

been in US jail.)

The last and most influential group is the Jewish billionaires. I say they are the most influential group because they were able to infiltrate most world governments and to act like a well-lubricated machine. For the moral and ethical standards of the Jewish group, you can draw conclusions from their actions. It is a well-established opinion that this group is behind the two world catastrophes, - the First and Second World Wars. At present, we have to realize that this group owns the world media and the US Congress. Most western governments have been totally infiltrated and are controlled by the Jewish group. The evidence for this is the foreign policy of the US, which blindly serves the orders and interests of this group. The arrogant and cruel actions of the Jewish state towards the local Palestinian population have turned the Arab people into their enemies. Instead of fighting with the Arabs, Israel simply uses America to attack and destroy their enemies.

The official version for the invasion of Iraq was 'weapons of mass destruction.' The public believes that the reason for this war was oil. But they are wrong! The reality was that Saddam Hussein was a real opponent of the Israeli occupation of Palestinian land and did not recognize the Israeli State. Saddam was giving $40,000 to the families of suicide bombers to rebuild their houses destroyed by Israel. This! Not oil was the real reason the US went to war with Iraq, - to remove Saddam in favor of Israel. The US doesn't have to bomb a country to get their oil! They have enough sophisticated methods for the exploitation of nations.

A good example is Venezuela – rich in oil, but the poverty and economic crisis caused by the US is enormous, and soon Venezuela will succumb to the external pressure. The US is destroying Syria, not because Assad is bad but because Assad refuses to let Saudi Arabia pass a gas and oil pipeline through Syria. Assad also opposes and is very uncomfortable to the Jewish State, which also wants to keep Golan Heights.

I have a message to the Jewish people:

I know that there are many wonderful people who oppose violence and corruption. The first principle of ethics is to consider your own actions before you start blaming others. The historical fact that the Jewish nation has been persecuted expelled, and hated everywhere they have settled in the past means that there must be the reason for this. The preservation of one's national identity, uniqueness, and culture and helping each other is a good thing, but acting like an organized society against the interests of other people is a totally different thing. Simply, you have to treat the others the way, how you want to be treated! Even in the present time, the Jewish people cannot find a way to live in peace with the local population! Why? What is the reason?

The Jewish group owns and have absolute control over the Western media and Hollywood. This gives them significant leverage over their opponents. The tragedy is that the ethics and mentality of this dominant group went in a completely wrong direction. They cannot understand that repression is the wrong way of power. Repression always creates a response that has to be overcome with increased cruelty and more repressions. This is a continuous circle of violence, where the end is public disobedience, blood, and revolution. - This is not a sustainable way!

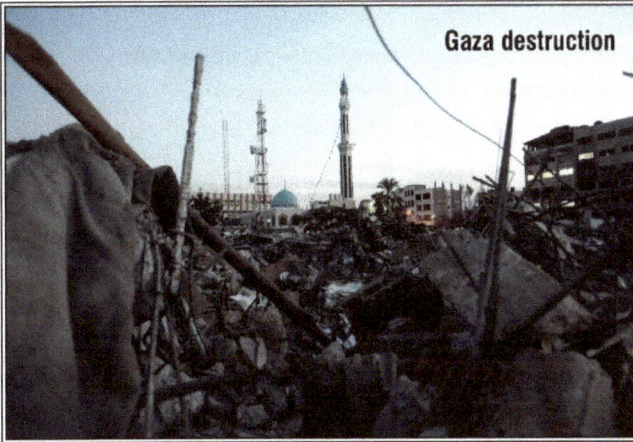
Gaza destruction

How the Jewish state treats the local population

The super-rich starts believing that they are super-intelligent to such an extent that they can do anything, disobey the basic principles of life and get away with it. This is neither smart nor intelligent, but simply arrogant and extremely stupid and dangerous for all of us. Do they really believe that they can make more money from bankrupt countries than from prosperous ones? Why do they transfer the Jewish-Arab problem onto the European nations and set the Muslims against Christians?

Israeli wall

Is this situation really smart, or bad and incredibly stupid? Intelligent people always have an elegant solution.

They oppose the Berlin Wall and the Jewish Holocaust, but they are doing the same things to the local population! Is this smart, or bad and stupid? I can tell them only one thing – intelligent people always act intelligently! Brutality and violence are the tools of primitive-minded people. In the chaos, wars, misery corruption, and uncertainty, I cannot see any intelligence, just real stupidity, primitive brutality, and a lack of any intelligent plan and vision!

Here is a simple example. - If you have a dog and you are good and encourage it, the dog will obey you and do anything you ask him to do, but if you are cruel, forcing and beating him, the dog will disobey you and someday will attack you and bite you. The situation is the same as if you start lying to somebody on a regular basis. They could believe you for some time, but when they realize that you are lying continuously, they will not believe anything, even when you are telling them the truth! This is the same situation, which we are having with information that comes from authority. The people become so disillusioned to the extent where they are ready to believe any nonsense that comes from independent sources of information. Afresh example of this is when many people started believing in the nonsense of the 'Flat Earth' theory.

What's wrong with our society? - Many things! We managed to destroy the foundation of our society – (the family). We inserted there the principles of the crooks –'irresponsibility and unaccountability'. After that, we took away the rights of parents to discipline children, without giving them any other methods or effective measures to replace this unpleasant, unwanted, but sometimes really necessary last resort. After that, under the cover of slogans for children's rights and the protection of motherhood, we inserted there unaccountability and reward for the irresponsible not loyal partner and parent with the adoption of the practice whereby, in 95% of divorces, the family house and children are given to the woman without questioning who is responsible for breaking the family. This is cruel, because millions of good, faithful, and responsible partners are punished, stripped of their assets, and their children are given to the unethical irresponsible partner. At the base of this practice are embedded, extreme cruelty, tolerance, and the promotion of injustice and immorality. Children need both parents, need love and peace to be able to grow into normal, ethical, responsible, and good people.

The younger generation ends up in a situation where unaccountability, irresponsibility, immorality, and violence are their everyday reality, and good examples of ethical and moral standards are absent in their lives. Education is stripped of subjects like history, the principles of democracy, social values, behavior, and morals. The movie industry is bombarding the young generation with constant killings, revenge, and senseless violence. Public ethics, morals, and a sense of right and wrong are effectively destroyed. The governments are banning tobacco, but legalizing marijuana! The people reluctantly accept as normal the examples of the wrongdoing of their leaders. The leaders must be role models for intellect, ethics, morals, and trustworthiness for the nation. - The acceptance of corruption reaches to such an extent that, when Bill Clinton was caught cheating on his wife and lied to the nation and yet remained in power! Ethical and moral people will not accept such leaders. The unaccountability went to such extent, that when George Bush Jnr. and Tony Blair lied in the front of the entire world that Saddam Hussein had weapons of mass destruction, and on the basis of this lie, they dragged their countries into a war that killed 1.5 million people, destroyed a country and they got away with this! Here is the irony of the next one. - Based on Obama's election promises and lies, he was awarded the Nobel Prize for Peace – he didn't do anything. The mere fact that he didn't even close the illegal Guantanamo Bay prison is enough to put him in the category of a person committing crimes against humanity.

I will explain this: A basic human right is that everybody is innocent until found guilty. Obama violated this basic human right by keeping in prison people without evidence, charges, and trial for ten years. By any legal and ethical standards, a person who violates human rights to such an extent commits a crime against humanity! The other disturbing fact is that the Taliban stopped drug production, but the US army in Afghanistan is effectively protecting drug production and poisoning the world and its own country. (Follow the money!) How have we come to such an absurd situation, that we start giving Nobel Prizes to such individuals? The US army is still in Iraq and Afghanistan; he created ISIS - (according to Ron Paul - US presidential nominee) and destroyed Libya, Syria, and Ukraine, is involved in the war in Yemen and is ready to start even nuclear war with Russia. Do these actions really deserve the Nobel Peace Prize?

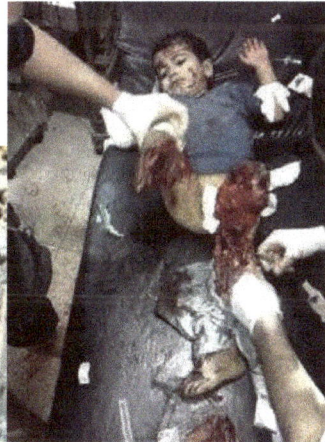

We give medals to those who commit such immoral acts!

I regret to tell you that the people responsible for the actions of the leaders are the people who tolerate them, vote for them and do not want to be bothered to hear or learn the truth of their actions. The irony is that the most of the Western governments now pass laws which are labeled 'anti-terrorist' that take away the first and fundamental principle of democracy – that the law must guarantee the freedom (or presumption of innocents) of the citizens. With one stroke of a pen, governments have abolished the basic foundation of democracy and given themselves the right to arrest anybody, imprison them for an unlimited period without charges or court interference, and nobody can even ask for more information. Isn't this situation the same as the one at the Guantanamo base? Unfortunately, in this case, you, my friend, you are the target! Doesn't this ring a bell? Doesn't anybody realize that these legislations are going in parallel with the building of new prisons, which exceed many times the capability required for the ordinary crime rate? The public does not understand what really is behind the current military experiments with unmanned drones. The people do not realize that the elite i perfecting the capability to be able to identify, target, and eliminate from any distance - eliminate anybody anywhere! They have face-recognition cameras for these drones, a capability combined with a total unaccountability and immunity of the person who presses the trigger and is hidden behind secrecy curtain of the corrupt government. - From the comfort of their office, some ruthless and crooked-minded people will have the luxury and ability to kill anybody, and no law, no responsibility, no court or public inquiry will be able to make them accountable. 'They' are perfecting a system of hidden and unaccountable terror! – Preparing system with the right to kill or imprison anybody who opposes them!

Another example of extreme public ignorance is the fact that when some Muslim extremists' detonated bombs in Western countries, people there asked in disbelief: why do those people hate us and want to kill us? We are peace-loving people!– Do those 'peace-loving people' know that their military has been bombing and killing the citizens of Arab countries systematically and methodically for decades? Or are those 'peace-loving people' ignorant of other people's life and sufferings? Do those 'peace-loving people' not understand the reality that the terrorists see them as the people who sent the army into their motherland and killed their family and children? - This is an example of extreme public blindness, arrogance, ignorance, and mass stupidity. And all this comes from a simple reason – the information chaos is working! It is blinding and brainwashing the public to such an extent that they don't want even to hear and to accept the truth!

If we really want to change the world for the better, we have to be able to accept the truth, even when it is unpleasant. That means that we have to be able to evaluate honestly and consider the facts. Not many people know that they are deceiving themselves on a regular daily basis. Usually, people develop the mentality to deceive themselves when they start to believe that the actions they take are always correct, and they are always right, their actions are justified, and that others are wrong. It is well known that this is the philosophy of losers – to blame the others for their own mistakes. And in reality, we are the losers because we allow the elite to deceive and exploit us by stealth and deception. - This is the real tragedy of our society – we blame everything and everybody else for the wrong things we do ourselves!

The situation is the same as many people who believe in God. They convince themselves to believe in God for their own personal benefit - to ensure eternal life for them, but in everyday life, they do not follow the ethical and moral principles of the religion of their choice. To be able to understand the world, we first have to be able to judge ourselves and be able to accept the truth, no matter how uncomfortable it is. The next step is to start learning the truth of the world we are living in. To find the truth, we always have to listen to the two opposing sides of the story, and the entire story, not only the official news where the facts are hidden, selected and presented with predetermined content.

The same criteria for truth and fairness we have to apply to the actions of our countries. We have to question the economic, political, and military actions of our countries. Are they correct and justified? Is our military allocated in our country, or are they serving outside? What is the purpose of our army? - To defend our country or invade and destroy others? The basic necessity for us is to be able to accept the truth for our actions and the actions of our society. Only when we are sure that we are correct, we can start judging others and ask them to be good and friendly. When we find out the truth about the world, our first duty is to teach our children and our friends the forgotten values and ethical principles of civilized people, and the responsibility for each of us not to tolerate lies, dishonesty, corruption, greed, and violence.

Greed and cruelty are the primitive instinct for survival of the animals. Ethics starts there, where the intelligence prevails over primitive instincts! We have to rise above that level and realize, that the only possible fundament of a civilized, intelligent society can be highly ethical and moral standards, and we have to start building it with consideration of our personal actions and values. If we decide to be good, it will be because we believe that we are good, intelligent and civilized! Not because we are afraid of God's anger or punishment. If we accept this strategy, we really will be good and ethical, and we will be able to do something to make a better future for our kids and for all of us.

To decide to be a good and fair person is only the first step, but it is not enough. To find out the truth and what is wrong with our society is necessary, but this also is not enough. This is actually the minimum value and understanding we have to possess to be able to choose the correct side and actions. Yes, action, because just to criticize and blame the crooks or make a demonstration might be OK, but it will not make any difference. The system of corruption is a well-organized system with the ability and tools to stay in power, and the actions of small groups of individuals cannot change this system. History is full of examples and attempts to fight against this corrupt system, but the attempts always fail. Only a very well organized social system, (not individuals) based on the total support of all people, can have a chance of succeeding, and not any system, but a system designed to fight corruption. The model of this system is embedded in the physical laws and properties and the organization of our universe! It is this idea of the system that was immediately suppressed when scientists discovered in the early 20th century that those principles are incorporated into the fabric of matter and the universe.

<u>A system of freedom, freedom of choice, self-determination and ethics, which incorporates measures for protection against corrupt and destructive actions, a system with unbreakable principles against corruption!</u> – This is the only possible system for our future! A system which is proven by the existence, harmony, and balance of our universe! The creator of the universe strictly obeys His principles for not intervening but despite this, is still able to help and give us the model to follow – the model of how to build our society, the model, which we really need to be able to prosper and survive!

The model of the system is in the following chapters. It is very short and simple, but the content and meaning are enormous! In a few pages are synthesized more truth and logic than we have in all philosophical books in the world.

I would suggest that you read it slowly and consider the meaning of every sentence. Then leave it for a few days and read it again one more time. Then you will find the reason why I suggest you do this. I wish you luck and success in your journey towards understanding the only possible way of how to build our society and our future.

THE SELF-DESTRUCTIVE ARROGANCE OF THE UNOPPOSED POWER AND THEIR ROLE IN THE COLLAPSE OF CIVILISATIONS

We, humanity are at a crossroads, where one wrong decision of our leaders can unleash grave disastrous consequences for the future and survival of our civilization. But the big question is how much we can rely on the intelligent decisions of our leaders?

It is well known that all the governments are formed mainly from politicians without a scientific background. This is the first worrying fact for their ability to have a credible opinion and correct judgment. Every government employs and relies on advisors in any aspect of their work. The problem is that the advisors are politically selected not always are the most knowledgeable and intelligent people in the area of their specialty, and the politicians are arrogant enough not to take any advice! One joke, which reflects the truth is that the government always uses scientific advice, just they select the scientist, who supports their own view!

Let see how the incorrect and wrong government decisions in the past have bitter and disastrous consequences for their nations and countries.

I would like to start with the above sentence, that any unopposed power becomes arrogant and disconnected from reality. This is the common sickness of any unopposed power which sooner or later starts believing that they possess supreme intelligence and can judge better than anybody else.

I would like to give a few examples from history. The well-known fact is that the destruction of the Roman Empire came from inside! The Roman aristocracy becomes incredibly corrupt and started fighting each other for supremacy and wealth. The leadership of the empire becomes dysfunctional and unable to make correct decisions, and this leads to the destruction of the sole superpower of the world!

The road of humanity never been smooth. From history, we know how many great civilizations have collapsed. The historians gave us different explanations, but the main common reason for the collapse of these civilizations remains the same. The historians are blaming natural disasters, epidemics, foreign invasions, but the truth is different from popular belief. Well organized society is able to survive any of the mentioned disasters and to repel a foreign invasion. There is one common reason for the collapse of all civilizations! – And the reason is the dysfunction and disability of the ruling institution of this society to take the necessary, appropriate decisions!

I will go back in history to give examples of such events:

I will start with the legend of the Tower of Babel.

The legend tells us that people become arrogant and want to build a tower to reach heaven! God gets upset and mixes their languages. The lack of coordination between the different groups leads to the collapse of their society!

Is more than obvious that the ancient philosophers have known that the organization keeping society prosperous, and the arrogance of the ruling people making the governing institution dysfunctional and this will lead to a collapse of their society!

We, the present generation, have built our civilization on the base of the inherited knowledge from our ancestors, and we have no right to be arrogant and dismiss such great intellectual messages coming from the past! This story clearly tells us that the ancient philosophers understood that the arrogance of the ruling institution of the society leads to their malfunctioning and the collapse of society!

We are facing the same pattern of events in the known history. The brutality of the Mayan elite makes the people very reluctant to defend them. The arrogance and dysfunction of the Mayan authority let 300 Spanish conquistadores conquer the whole country!

The same thing happened with the Aztecs; the governing apparatus has turned itself into one cruel, dysfunctional institution, hunger for wars, and human sacrifices. The advanced and well-organized civilization of the Aztecs is more than capable of defending themselves against a few foreign ships, but the dysfunction of the authority makes them easy prey!

We can see the same pattern of civilization collapse with Egypt. This great and powerful civilization has succumbed to Roman pressure, because Egyptian authority and Pharaohs instead of building their country, in pursuit of personal glory and eternity they have bankrupted the country with useless projects such as the great pyramids and colossal monuments. The pursuit of personal glory has disconnected them from reality. Instead of fighting and defending the Country, Cleopatra decided to sleep with the enemies! I am sorry, but this is not the best decision for the leader of a great Country.

We have mentioned above, that the collapse of the Roman Empire came from the same source - dysfunction of governing structure.

The next example is the British Empire! It was the most powerful economic and military machine in the world. There was no any reason for it not to remain and preserve its supremacy to the present day. Again, the British Empire was not defeated! The Empire crumbled as a consequence of the wrong and incorrect decisions made by their arrogant and disconnected from reality leaders!

The greed and arrogance of the decision-makers of the British Empire led to the two Great World Wars! - Instead of co-operating and benefiting from the emerging German industrial mind, the British Empire decided to destroy Germany by war. Yes, they succeeded in doing this, but in the process of two major wars, the British Empire has bankrupted itself. All the wealth of the British Empire and European nations went to America, where the river of gold and money create conditions for the emergence of a new world superpower.

The next well-known example is the rise and fall of Hitler. Hitler successfully built the most powerful army and industry in the world. In practice, if he has been intelligent enough, now the first language in the world would be German!

Hitler was writing that the reason for the German defeat in the First World War has been the fact that Germany was forced to fight on two fronts. Despite this knowledge, the arrogance of his status as the unopposed supreme leader of his nation led him to believe in his supreme intelligence, and his ability to deal with any situation! When his army was able to defeat Poland and France in a matter of weeks, drunk with success, Hitler was blinded to the extent, where he ignored the lessons from the past and the advice of his generals not to open a second front! Hitler has been in an undefeatable position, because he controls the European industry, and had the natural resources of allied USSR. USA has been neutral, France defeated, and England had their supplies cut off. In a situation like that sooner or later, England would have to start negotiation for a peaceful solution because they have no prospect of winning the war! For losing the war, Hitler can blame only his arrogance!

Recently we have been observing the self-destruction of the Soviet Union Empire. The empire has all the necessary structures to remain world power –they had the strongest army in the world, producing 30% of world GDP, had the strongest nuclear arsenal, and one of the best education and scientific research facilities in the world! But this empire has only one problem! – The authority has been unchallenged for a long period and becomes a selfish arrogant institution, which doesn't care for the needs of their own people. When the time came, the people with enthusiasm destroyed the empire, which was tyrannical to them. The Soviet Empire wasn't defeated! They fell victim to the arrogant and dysfunctional behavior of their ruling elite!

The next example of wrong political decisions, which destroying a world superpower, is current. - We, the present generation, have witnessed the self-destructive actions of the world superpower – the USA. You would ask what I mean? I mean that the USA is using the same incorrect methods and heading in the same wrong direction as the British Empire. The USA has the best industry in the world, the USA has the most advanced research centers, which was producing the best-innovated technologies, which was assurance, that the US industrial supremacy would be unsurpassed and unchallenged in the near future!

So... what had happened? The arrogance of US authority as an unopposed and unchallenged power again led them to become disconnected from reality, and they started making wrong fundamental decisions. The financial and industrial elite of the USA were not satisfied with the good and steady profit from the American economy and decided to make even more money! With a simple stroke of the pen, the unwise greedy elite has exported American industry in China with the aim of a fast-enormous profit!

OK, this was working for a while, and the rivers of money-filled their coffers but at what cost and consequences? Despite their thousands of economic advisors, the corrupted elite fail to understand simple facts! - You cannot make more money from a bankrupt country than from a prosperous one! The empires are keeping their army and bases in order to ensure their privileges to extract natural resources from the colonies and to produce and sell their industrial products to the world. The US elite does not understand this simple principle, that if they destroy the US economy, it will be pointless to keep their military bases in the world - they just will bankrupt their own country!

Currently US elite is struggling to patch up the situation with near-zero interest rates, and with endless printing of new money, with local wars for petrol reserves and revenue, and with hope, that the military production can sustain US economy! This is just an empty illusion! The vacuum of manufacturing is exhausting the financial reserves of the American population. The US infrastructure is crumbling; there are not enough funds for medical and social programs, for education and scientific researches. This situation is irreversible! There is only one way to reset the situation! – New World War! But the problem is there - we already will have this war, but such war in a nuclear age is unwinnable, and there is no guarantee that even the elite will survive!

And now my friends, I would like to explain to you the best method which will give you the ability to find out what somebody's real intentions are, regardless of whether it is an individual or institution! - I call this method:

Facts, Logic, and Intentions!

In any aspect of life and science, the most important elements of truth are the facts! And the facts, not the declarations, rhetoric's, promises, and official announcements are the real indications what a person, politician or government is planning to do! I am giving you this knowledge with the aim of

making you aware of what's going on around you and on the international arena to be able to make your correct judgment of the current situation.
We have to realize that despite the declarations of our leaders they usually are planning different things, and the most important realization for us is that the government could be absolutely wrong and unintentionally can destroy the country more effectively, than any external threat. We know this because this had happened repeatedly in the past and even in the present day we are witnessing incompetent leadership of the US, which is draining the country resources to maintain needlessly enormous military machines!
The realization that we cannot rely on the competence of our governments is putting urgency to do something, because the humanity is on a crossroad, and we have to make sure, that we will take the correct road for our future!

WHAT IS THE REASON FOR THE SOCIAL REVOLUTIONS?

We have to learn the lessons from history in order not to repeat them.
Many people are repeating this sentence, but to be able to avoid disasters, we really have to know the reasons why these disasters had to happen! It will be really helpful to know what was the reason for the social revolutions of the past is.
It is simple, - the social system at the time fails behind and hasn't been able to reflect and serve the reality of life and the necessity of the technological advancements of the society! When the old social system becomes a significant obstacle for the advancement of society, it inevitably is bringing social revolutions! It is very important to know that the social system and structure of our society is not determined by individuals, nor by groups, or ideology. - The truth is simple: The social system is determined by the necessity of the way how society produces and exchanges goods and values! - This simple fact is well known but our politicians like to have the upper hand and this is the problem!
I will provide a few examples from history, where every economic era ended and changed its social system by violence and revolutions:

- In the stone age era was no production, the tools and food have been limited, and for the group to survive, they must share food and tools – in that time everything has been own collectively and the primitive society obeying this rule has managed to survive the most severe climate events.
- When agriculture brings prosperity, and the humans have been the sole producers of the goods – they, the humans become subject of ownership of wealthy individuals, and the era of slavery has been established.
- The Slavery slowly becomes an obstacle for progress, because the slaves have no any incentives to work hard, and produce more, but their owners must provide them and their families with everything

they need for living. This inefficiency of slavery has been succeeded by the next more progressive era of Feudalism.

- Feudalism gives limited freedom and financial incentives to the farmers to produce more and reducing the liability of the 'Owner.' On it's our term, the Feudalists social structure also becomes an obstacle for human progress when the machinery has been invented, and the industry needs free workers to function and prosper.
- The Capitalists era was a more progressive social structure because of the way of production involving machinery and free human operator. The social structure brings financial incentives to both sides - to the owners and the workers to compete for sophistication! It does bring to humanity the technological revolution. The Capitalist social structure is based on trade and exchange of labor for goods and money between the "Owners" and labor forces (or the majority). But in the way how all previous social eras have exhausted their means. Capitalism also is not immune to this process and gradually becomes an obstacle for the next more progressive social structure.

The new way of production and distribution of goods will determine when Capitalism will become a total obstacle and have to be replaced with the next more progressive and appropriate for the new reality social structure.

The next more sophisticated way of producing our goods is the new era of computerized production lines not involving many people, but producing a vast amount of goods. At the moment we are exactly on this stage of technical and economic development!

What new social structure we will need to satisfy the demand for the new way of producing and exchanging goods and money?

Capitalism has exhausted its ideas for social progress and become an obstacle for human development and progress. The new way of producing the goods makes the social structure of Capitalism absolutely inappropriate and out of touch with the new reality, because the basic idea of Capitalism is exchange of labor, goods, and money between the 'Owners' and labours (the majority), but in the new emerging reality of life this social structure becomes senseless because the job for the working people becomes less and less available and they are losing their normal way of income, losing their buying power and the circle of exchanging goods, labour and money stops!

The irony of the new situation is that the 'wealthy' - (the top 1%) can produce the goods without the help of the people, they can make and accumulate a lot of money, but the problem is that the Capitalist circle of exchange is broken! The circulation is ceasing! The social structure of society is going to stagnation and is not reflecting the new reality! The people gradually are losing their jobs and income. The increasing size of the services sector is not a productive activity and cannot stimulate the economy! It is just consuming, nothing else! This is the key crucial factor! - The new sophisticated way of producing the goods need new social structure of values exchange to meet the demand of the new reality of the life because the current system brings stagnation, where the rich are accumulating the money, but they cannot release them back in circulation, because the people can offer them nothing in exchange! - The people are losing their jobs, losing their income, and buying power!

To destroy Countries and go and rebuild them after is not a solution to the problem! It is a problem by itself! - This cannot continue forever; it cannot create sustainable jobs and cannot solve the emerging financial and social stagnation! To print more money is not a solution either! This stagnation will become deeper and deeper and inevitably will lead to a social revolution how it happened in the past! <u>We need the understanding that not the individuals or political movements will bring the social revolution!</u> – <u>The financial stagnation and the inability of the current social system to meet the requirements of the new reality will bring it!</u>

This is the advice which I would like to give to the Elite:

Please, look back in history and don't repeat the mistakes of the past, because they are covered with the blood and misery of millions of people for no reason! There are two ways of social transformation: Controlled or Spontaneous.

In our history, we have mainly uncontrolled Spontaneous social changes, made by uprising, wars, and revolutions. There are a few examples:

Spartacus made a mortal blow to Roman slavery; The English Civil War ended the Feudalism with the Tenures Abolition Act, The French Revolution abolished the European Feudalism, the 1917 Bolshevik revolution ended the remaining Feudalism in Russia and American civil war liberated the slaves and the work force for the emerging industry.

All these changes have been dictated from the necessity of a new social structure! Not by the individuals, Royalties, religious or political ideas!

To suppress the emerging social changes is the same as blocking the relieve valve of a boiling pressure cooker! - It will blow violently! And in the following chaos, always has emerged violent groups and individuals like Spartacus, Cromwell, Napoleon, Lenin, Stalin, and also collectively organized terror like the terror of the French Revolution and Bolsheviks. <u>We are on the verge of such social changes, and suppressing them is not the best idea</u>, because this will give way to bloody uprisings, violence, wars, and political assassinations like the one of the king Charles 1; French king Luis 16; Russian Nicolas 2; Nicolae Ceausescu, or Gaddafi.

The super-rich Elite have to realize the bitter reality that they cannot continue making money from bankrupt countries, bankrupt governments, and increasingly jobless society! – The elite are accumulating only the fake value of the newly printed money from thin air! In the coming stagnation, the assets will lose value because they cannot make a profit and cannot be sold!

Capitalism has exhausted its ideas and become an obstacle to our future development. In the previous chapters, I have reviled the regressive methods of the Capitalists system to cling to power. It brings us the new 'Dark Age' in science, countless violence, destruction, and moral degradation. The most alarming feature of this system is the increasing imbalance between our technical advance and our knowledge, ethics, and morals.

Ethical standards are the only reliable tools for controlling dangerous technical and scientific inventions! – Ethics, consciousness, and emotions are the real difference between humans and intelligent machines! - AI will never possess those qualities like ethical standards – loyalty, trustworthiness, love. AI will always remain a dangerously efficient, ruthless, and emotionless machine with no loyalty to us! We have to realize that our emotions, ethical standards, and morals are our most valuable qualities, which we have to preserve at any cost because those qualities are the key to our collective survival!

It is time for reason and intelligence to prevail over greed and stupidity!

The Elite should give way, support and protect the intelligent and ethical individuals, who are slowly rising their ideas for the new and more progressive social system - system, which will reflect the necessity driven by our social development. This will be a smart approach from the Elite, because the ethical and intelligent intellectuals will not bring social instability and chaos, but will provide a smooth and elegant transition to the newly emerging social structure, which will ensure peace, prosperity, stability and the future of humanity.

For two thousand years, people have continuously tried to find and elected a good leader with an honest team who will work for the people's interests. The tragedy is that when the new leader sits on the throne, he forgets his promises and starts enriching himself and his entourage. This scenario is repeated over and over the centuries, and nothing has changed. It is more than obvious that this approach is not good and does not work, because this system of governing is designed for the minority and the corruption always to succeed. The hard to detect imbedded principle of corruption is that the elected representatives have total immunity and unaccountability which is giving them the freedom to do anything they want! The immunity and unaccountability of the governments really mean that they are not legally bound to serve the people and to follow their election promises.

There are countless philosophies and models for governing and social structures, but all of them are in the same category – just utopia! That means that they are just very nice and good wishes, but are neither practical nor workable because all of them are based on the same wrong principle – to elect an "honest" government without including there a legal system to control the government actions. - This is the flaw of all utopias!

The super-rich elite for two millenniums has perfected a system of hidden legalized corruption embedded in our governments. They realized that when they invest money in industry, the return is not secure, and is not more than 20%, but when they "invest", (or bribe politicians), the return is absolutely secure and is 10 times greater than the spent money – the profit from political favor is on average 10,000%. - In this scenario, you must be very naïve to seek to elect honest people as they inevitably will be bribed or blackmailed. -The banks say that the opportunity makes the thief! - And this is the bitter reality of life, because when loopholes and opportunities for corruption are embedded in all structures of the current systems we inevitably will get corruption! - To avoid this, <u>we must implement a legal social system where politicians will not have any opportunity for political favoritism or wrongdoing</u>. - This is the only way to make a good and fair governing system. These measures are not utopia, and is really simple, achievable, and effective! Please read the following chapters where is the detail explanation of how this can be achieved.

MYTHS, LIES, AND ILLUSIONS

The popular belief that we are living in democratic countries is well established. Politicians have succeeded in convincing the people that this is "the best democratic system" that we can have and that "they", - the politicians, are the servants of people and democracy. The problem is that people really don't know what the fundamental difference between republic and democracy is!

The majority of people believe that the republic is a real system of democracy, and we are living in the "best democratic system", but this is not correct! First, what is a democracy? - Democracy is the rule of the majority – (the people).

In the current system of Republic, the people have to surrender their rights of governing to an elected minority, <u>and there is no any legal guarantee</u> that this minority will rule in the people's interest! In practice, we end up in a situation where **the elected group of an immune and unaccountable minority is ruling over the majority.** - <u>This is the exact opposite of the principle of democracy</u>!

The second point is: Do the politicians serve the truth and people's interests? It Is well known that politicians are masters of politicizing the truth. In real terms, that means that they are masters in twisting the truth by presenting only the facts which serve their agenda and purpose. They are able to convince the people in different countries to believe in exactly the opposite things for the same events. - That's how they are able to put nations into conflict and make wars on the base of lies. In reality, a gifted politician is regarded as the person who can twist the truth in the most elegant way to serve his agenda.

By any moral standard, a person who twists the truth is not an honest person. <u>And in real terms, this effectively is to appoint professional liars to rule the country!</u> – Is this really being "the best democratic system" which we can have? Just think about the absurdity of the situation and the wrong social structure we have to create and tolerate!

<u>In reality, the people end up with only the right to replace one crook with another!</u> - This is not democracy and is not the best system by any means! The reluctant acceptance of this tragic situation is a result of misunderstanding the principles of democracy and this misunderstanding is the major obstacle for the establishment of a real democratic system!- This

situation has to be corrected!

The public has to be informed and has to develop an awareness of the misinformation which they are bombarded with, and understand that this misinformation serves the minority interests only!

In the following parts, in short, and clear terms, I will describe what are the major problems with the current governing structures and how easy it would be to correct them to establish democracy. - And not just any democracy, but real democracy! Stable and protected from corruption democracy!

CURRENT MODEL OF DEMOCRACY AND ITS FOUR MAJOR PROBLEMS

The current model of democracy always stops short of completion at the point where the politicians are democratically elected with the assumption that they are very honest and will serve only the interests of the people. - This assumption of honesty continuously and repeatedly is proven wrong, because politicians are not really the most honest part of the society and the system gives them total immunity and unaccountability. So, if you cannot trust somebody, you must be able to control him! - **This is exactly the biggest fundamental problem of the current model of all present so-called "democracies" – a lack of public accountability over the most important institution in every country - the government.**

In practice, this is a wide-open door for abuse and corruption **because the people have no legal system and legal ability to control or influence any political decision.**

We have implemented law requirements in any aspect of our life, but in the most important institution, we are failed to insert any measures or systems of accountability! Why?

The second fundamental flaw of the current so-called "democracies" is the fact that in reality, this is not a democracy, but is actual and effective rule of an elected and mostly corrupt minority over the majority, - (the people)!

The third and fourth problems are:

a) That the two parliamentary bodies – the Senate and House of Representatives (Parliament) – are represented by the same parties, and this makes them unable to control each other.

b) The ministerial portfolios are taken by politicians, who are just politicians; - neither qualified nor specialists in the area of their duties.

In short: - The current model of so-called "democracy" allows all political decisions to be made by a small group of unqualified people protected by immunity acting in self-interest or for political reasons only! - This is not a governing system designed to work in the people's interest at all! - And is not a democracy!

We have very strict criteria for qualifications, capabilities, and knowledge in all aspects of our society, but we fail to insert any of these criteria in our most important institution of the country, - the Government!

The current model of governing is exactly the opposite of the fundamental principles of democracy because **democracy is the rule of the people**, is not rule *over* the people! - This is the current absurd situation, where the people have no right to govern, and where this system is constantly labeled and presented as "the best democracy which we can have". - This is a tragedy; it's not a democracy!

As a result of the incompetence and unaccountability of governments, there is undeniable evidence of wild corruption, economic mismanagement, social degradation, and self-destruction of nations. It is time for this situation to be corrected because any authority left without control gradually loses direction and becomes corrupt, ineffective, and a totalitarian oligarchic egotistical system. - And this system definitely is not the best system that we could have! This system brings all the tragedy to humanity - wars, recession's, poverty, ethnic tensions, inequality, and misery!

Can you really believe that nothing can be better than this corrupt system? What would be wrong if the people have legal rights and established a legal system to approve or disapprove any proposed legislation, measures or deals by the government? - This would-be democracy, where crook individuals will not have the power to bring us wars, recessions, and destruction! Why do we not choose to have real democracy? - It is not too difficult for this to be done!

THE MISSING PART OF THE DEMOCRACY AND HOW TO RESTORE IT

The dream for democracy, equality, and justice is as old as human civilization. Countless different social organizations and structures were tried, but good and lasting democracy has never been achieved. Lessons from history show that every democratic system sooner or later is hijacked by internal corruption that weakens the system and gradually destroys it. Examples start with the biblical story of Babylon, followed by the crash of Greek civilization, the Roman Empire, Egypt, the French Revolution and on...and on. Even such organizations like the Catholic Church, which is supposed to be based on love, generosity and self-sacrificing, became totalitarian and in the Middle Ages started torturing and burning books and opponents. There was an attempt to create the perfect society by implementing the communist system – where everybody is equal – but even this attempt was hijacked and finished like a system where some people were more "equal" than the others.

The most recent example is the decline of the Western World where the system of free enterprise and competition for the benefit of the society has become something very different from its original idea because the corporations ended the free competition with monopoly, which controls governments and delivers most of the benefit to 1% of the population!

Western governments have allowed big business gradually to relocate the industry in the developing world in pursuit of big and fast profits. We witness the decline in cultural and moral values, educational standards, and economic collapses where unsocial governments are favoring big business and gradually reducing vital social structures like pensions, health, education, and science. Western countries end up in a situation where people have no money, businesses are struggling and have no money, the banks have no money, and governments have enormous deficits. Only some mysterious private figures manage to collect all the money in the world and call themselves "legitimate" organizations like IMF, Federal Reserve, etc. They are powerful enough to be able to influence the media and government decisions and policies. The signs of self-destruction of the western world are undeniable. <u>Societies, where the governing institutions are losing the ability to serve the interests of the nations, have no future.</u> Sooner or later, they will collapse under internal and external pressures as it has always happened in the past.

High-level corruption is the most difficult thing to combat and control because these officials are managing all governing institutions, media, and institutions for public control. The two-party system is a joke – like two hands of the same body – <u>and people have no any legal way to influence any government action or decision.</u> – And this is wrong, absolutely wrong! - Very, very wrong!

<u>This is the missing component of the democracy</u>– **the people's ability legally and effectively to be able to approve or disapprove any law, any decision, any measure, or policy!** - That's why always any attempt for establishing democracy has always failed **because effective public control has never been implemented over the decision-making bodies! -** Any authority without control sooner or later is destined to become arrogant, corrupt, and totalitarian. The control of government decisions must always be in the people's hands!

To change this situation, it is necessary to inform the people and explain to them the main reason why democracy has never been established and how to change the constitution to be able to be in control of the government and how the new democratic system to be corruption-protected! -

To be able to be in control, once a year the people have to be able legally to vote and approve or disapprove any new proposed legislation, measure or deal before those legislations can be implemented and become law. The people must have active and continuous legalized control over the government to prevent corruption from taking over again!

Have to make politicians accountable and responsible for their actions! There must be implemented a system for financial benefit or benefit reduction for each of their parliamentary voting in respect to which way they vote, - either for or against popular or unpopular legislation. Any politician who votes more than 40% against the peoples will automatically lose his seat and benefits at the time of the yearly legislative approval peoples voting. The family businesses, bank accounts, and assets of any person involved in the legislative decision-making process, such as a government advisor and data provider, must be publicly transparent. - This accountability measure will not give politicians the chance to become corrupt. They will run the administration, make proposals, but the decision to be implemented will be made by the people! – This is a democracy!

To enforce and protect the law and democracy, the Legal system of the Country also must be in public hands and control, and the same measures of public control must apply over the appointment and accountability of the Judges. Judges must be public accountable! This will guarantee that the law and Constitution will be legally protected from politicians and corruption.

HOW TO BUILD STABLE AND SUSTAINABLE DEMOCRACY

The people in the Western countries are fed up with their unsocial governments, but they really don't know how to change the situation, because to exchange one politician with another or exchange one party with another party is not working. It is necessary to change the political system and give legal power to the people to have the rights to control the fairness of every political decision before these decisions are implemented and become law. This can be done by a referendum to change the constitution.

To be able to succeed such referendum, we first have to establish a political movement and party that will promote the idea for real and protected democracy. When this party wins a majority in parliament, - only then can such a referendum could succeed! Good work has to be done to educate the people and promote the new system to gain popularity and public support.

The new governing structure:

We have to simplify the governing structure to make it work better.

President, or leader of the country is a very dangerous prospect to democracy. The sole leader is a potential tyrant, and he inevitably will interfere, politicize, and will disrupt the work and democratic principle of parliament. Democracy doesn't need tyrants!

If the Country needs to negotiate with another country, just have to send the appropriate minister there. For example, the Foreign minister can negotiate political issues. If the Country needs a trade agreement with another country, the minister of trade can do this job. We have ministerial portfolios for every aspect and issue. We just have to let them do their job without pressure and interference!

Senate's function for approving deals projects, budgets, policy, and virtually everything will be taken by the people. (Senate must be abolished). To be ineffective control, the people have to vote once a year to approve or disapprove of the work done by each ministry, government institution, and any proposed legislation, deals, measures, and decision. Only after the people's approval of the approved projects, deals, and legislation will be implemented!

Some people are afraid that the parliament is passing too many legislations to be able to keep track of them. Currently, the legislative system of our countries is bombarding us with an avalanche of new legislation in the rate of thousands per year which are written on complicated language. These legislations can be interpreted in many different ways, and are on language, which only the lawyers can understand. - Such kind of law, which nobody can keep track of it, and only a few can understand is serving to produce only chaos and confusion!

The law of the country must be simple and clear! The Law must be formulated on correctly selected ethical principles and to be written on a simple, clear and easy to be understood language.! - Every single law must have a clear meaning which cannot be subject to different interpretations. Once the laws are established, the Laws must be permanent and not be changed! - Changing and replacement of the law must be an absolutely very rear event and to be done by the people through referendum, not by the politicians how is now! The Parliament will not discuss the law, the Parliament will run the country affairs only! - This is the job which the Parliament must do!

Stabile social system we can have only if we have simple, correct, clear, and unchangeable law and principles. Every ministry has to be free to do their work according to the adapted and approved plan and budget without any external pressure and influence. Every government institution must be accountable to the people and not to hide their activity behind secrecy!

The elected government representatives - (The Ministry) must be administrators only, which is running their portfolio by the proposed plan, budget, and time schedule.

The Parliament must be the assembly which discussing and coordinate the work of all ministry and deal with the current problems. The parliament will not have the rights to discuss and change the existing laws! Only the referendum can change laws!

We have to take a lesson from our surrounding:

The Universe is an incredibly complicated system - much, much more complicated than our society. The Universe is based on unchangeable law and unchangeable principles, and exactly this is giving its stability, balance, harmony, and eternal existence!

The correct chosen unchangeable Law The Universe is giving us a perfect example how we have to build our social structure! - <u>Chose the Law, locked it, and don't allow the politicians to play with!</u>

To satisfy the qualification criteria:

Before each general election for each ministerial position, there will be a public nomination where the scientists and specialist have to reveal their policy for the next four years of a parliamentary term, and then the people will have to vote and elect directly for this position the most appropriate person for each ministerial portfolio.

Every year before the compulsory annual voting, each minister must present a detail public record of what his ministry has done in the past year. If the people are not happy with his work and their approval is less than 50% This minister automatically losing his seat, and the next on the list of candidates for this position will take the position and continue the work for the next year. This will provide the ability to the people to have the legal power to demand the election platform be implemented correctly with the budget and on time! This will provide a legal system where well qualified and capable people will be elected for each position, and in this way, we will have effective and competent parliament that works in the people's interest.

The people's rights of control will be guarded by an effective constitutional legal system for control and protection of democracy in the hand of the people.

The long-promoted idea, that the people are stupid and must surrender their rights of governing to an elected and unaccountable minority, repeatedly has been proven wrong, because the immunity and lack of control allows the ruling minority to do anything they like, to politicize any issue and to twist the truth in one way or another. This political manipulations, incompetence, and corruption serve only the interest of the selected minority and bring to the world only endless wars, instability, inequality, poverty, and uncertainty. This has to be changed, and this is the most elegant and simple way to change it!

FUNDAMENTAL SOCIAL AND MORAL PRINCIPLES OF
MOVEMENT FOR PROTECTED DEMOCRACY

- The main principle is to establish fair social, moral, and ethical standards of civilized nations.
- To protect social harmony, personal rights, freedom of expression, religion, social equality and justice.
- To protect the rights of the family to raise and educate their children in an environment where violence, crime, pornography, and drugs are not allowed and are not the main focus of the entertainment industry and the media.
- To oppose aggression and war and classify them as crimes against humanity.
- To protect the rights of fair information, where the media must provide complete and full information - the truth, all the truth, and not only selected parts of the truth with pre-selected conclusions.
- To abolish secrecy and establish freedom of information and freedom of expression across all social organizations, government deals, negotiations, and policies, (excluding only the country's military

secrets and gathering foreign intelligence information).

This is a basic, minimum package of necessary ethical principles:

The purpose of the movement is:
To establish a constitutional system of corruption-protected direct democracy

The hierarchy of authority:
The supreme act of democracy is the referendum! Followed by the Constitution which is approved by referendum! The law of the Country must be clear defined and comply strictly with the Constitution!

Citizen's duty:
To express their will by participating in the elections! They must be well informed and make wise and educated choices!

Principles of the Protected Democracy Movement:

- The first principle is: To protect itself from internal corruption. — When the principles and constitution of the movement are established, no one will be allowed to discuss or offer any alteration or change to the principles of the movement. The principles are permanent and unchangeable! Otherwise, if changes are allowed, corruption will find a way gradually to modify, replace, and destroy the principles!
- The main principle: Promotion and establishment of a governing democratic system protected from internal corruption, where the people have constitutional rights and an established legal system to approve, disapprove or reject any government action, legislation deal, project or decision.
- Voting: Voting is the most important act of democracy and is a supreme responsibility of each person to the society. Voting must be compulsory and must be protected from corruption! Voting must be done with a physical ballot paper (ballot paper and retaining copy). The voting will be done by perforation of the ballot paper. Simultaneous perforation of the ballot paper and the retained copy will make any manipulations impossible. Each ballot paper must have a number and the perforated retain portion (copy) which will give ability each person to be able to prove how he was voting and to trace how his ballot paper is recorded in the system trough internet! The ballot papers must be safely stored for a certain period of years, or decades! — This will eliminate the possibility for manipulations, adding or hiding ballot papers or incorrect vote counting!
- Ethical standard: To establish legal foundations for an ethical, moral,

and civilized society, where the ethical and moral standards will apply to all aspects of life, to all institutions and government without exceptions.

- Family protection: To protect the family as the most important place in our society, where each of us grows up and develops as a person. The family is a lifelong commitment between a man and woman to help each other to provide a peaceful and moral environment for raising the future generations. Breaking this commitment should be regarded by the law and society as an intentionally harmful act, destroying the life of all family members.
- Pension age: To adjust the pension age to serve its purpose - the people to have a chance to retire when they get older, not to get the pension in the last year of their statistical age limit of life.
- Pension funds: Have to make pension funds tax-free, and government growth guarantee and preservation, with no administrative fees allowed to be applied. The funds must be used only for public infrastructure until they mature.
- National resources: The natural and mineral resources are a national treasure – that means that these resources belong equally to all people of the Country! - Nobody has the rights to own them! And nobody can exploit them without paying the royalties and dividends directly into people's bank accounts! - And only then, when this is done, the Government can apply income tax on the deposited money!
- Foreign policy: To establish an ethical and lawful standard in foreign policy, where double standards are not tolerated and where war, aggression, and violence are regarded as crimes against humanity.
- Finances: Every nation must own and control its currency and finances! Have to make monetary and financial reform and nationalize the Reserve Bank of the Country. Then the benefit of the Interest, Inflation, or printing new money will benefit the people of the Country! - (Not individuals how it is at the present)!
- Monetary policy: Currency not supported by value is prone to speculations and instability. To end financial uncertainty and provide stability, we have to return to the gold standard! – Only this can change the current system which allows bankers to produce money from nothing, lending them to governments and individuals and charging them interest! We have to make the share market a long-term investment platform only! Free of short selling and speculations.
- Justice: – Justice is the legal guard of society; it is the law enforcement instrument and is the guard of the principles of democracy. No matter how good and democratic laws the Country have, if these laws are not enforced strictly and swiftly, the corruption very quickly will take over! The appointment and control over the judges must be strictly in public hands, not in the hands of politicians! The Justice always must be

delivered and be proportional to the size of the offense! Forgiveness not serving justice! - It is a philosophy of the evil society! Forgiveness of crime is a crime by self, because It is the encouragement of, and cooperation with crime and is a way of tolerance of injustice, cruelty, and crime! – The law and the Judge have no right to forgive the crime The Law must deliver fair and proportional sentences to any committed offense no matter who is the offender!

- <u>To protect the rights of fair information</u>: The law must demand the media to provide complete and full information - the truth, all the truth, and not only selected parts of the truth with pre-selected conclusions. Secrecy In any institution or government activity or deals is not allowed!

- <u>Discrimination:</u> No one should be discriminated on the base on his ethnic, religious, or physical appearance. All people have to have the same opportunity and the same rights! - But social equality must not be mixed with the necessity of criteria of knowledge and capability for any particular position! The strict criteria of capability, responsibility, and ethics must apply especially to any leading position in society!

- <u>Gambling</u>: Have to stop gambling in any form, like sport, competitions or entertainment, because of its unsocial unproductive nature, where the naivety of people is used for the unfair financial gain of others.

- <u>Drug policy</u>: To have zero tolerance for drugs, with harsh penalties for selling drugs or promoting it! Have to implement a system to help the addicted and destroy the market and the financial incentives for drug to be sold. (There are already good examples of properly implemented such systems against drugs.)

<u>Business structure</u>: The current capitalist system has lost its direction and social values. To correct this situation, we have to make constitutional changes where the interests of the family and society have priority over the established laws and practice of making a profit from everything, regardless o the consequences for the society. That means that we have to create two social sectors – the non-business sector and the business sector.

The two-business sector structures would be as follows:

a) <u>The non-business sector</u> would have to cover family home services – water, electricity, gas, and the internet, as well as education, health, and pensions. These services must be provided at a basic cost by the government, and no one should be allowed to make a profit from these sectors off family home services.

The family protection measures have to apply in order to change the social philosophy and to serve the people and business fair and equal!

b) <u>The business sector</u>: It will cover all manufacturing industries, services, and virtually everything else where the non-business sector applies. The businesses will be allowed to compete and make profits under strict regulations with measures to prevent corruption and monopolization of the market. Have to encourage healthy competition and maintain good ethical, moral, and business standards.

The view that governments have to control the industry only by applying taxes is wrong and irresponsible. Governments must be in control of the economy and be able to provide affordable business infrastructure for the industry: roads, rail, ports, communications, energy, and finances by owning them, or at least being the leading company in each sector to provide a healthy standard. Only then, when we have stable and affordable government provided the business infrastructure we will have the foundation of certainty, stability, and sustainable and prosperous industry and ethical and free of corruption society.

Implementation of the fundamental principles: To implement and preserve the correct direction of the fundamental principles, the movement for protected democracy will not be allowed to formed any coalition or to make any policy compromises. When the government is formed, the movement first must implement the main principle of the movement - which is to give legal power to the people to approve or disapprove any legislation or measure, and then it will be able to implement gradually and carefully all the other fundamental principles of real, stable and sustainable democracy.

Solon, the father of Democracy

I like to say a few words at the end of this book. In this book, I have tried to give you the best explanation for the vital questions of life which I have found in my quest for knowledge and truth for the real structure and meaning of our world. I have tried to use the best scientific way to reville the truth and to act like an independent investigator using correct facts and logic and make sense of them. These findings are different from the official version presented by science and religion because I have considered all facts without preference and hidden agenda. With correct available facts, we can find the answer to the most important questions: Is there God? Is the Universe having a beginning or end? Where do we come from? That there is a similarity of the religious statement -that God creates man in his image and the findings of science because the universe poses the same building blocks like us - consciousness, information computing capability, intelligence and also have a physical part. The shocking realization of the available scientific facts is that the universe is an incredibly complex physical system that possesses all conditions, all elements, all quality and capability of a super-computer and super-intelligence, and there is no any reason not to act as one enormous intelligent living organism!

It is time to accept the realistic understanding and knowledge of our universe! It is time to accept the presence of universal formula, which applies to the entire universe, to every part of physical matter and every living organism! - The formula which the greatest scientists and philosophers of the past were seeking without result! - The formula, which the corrupted elite is hiding from us for a century! It is simple, as every ingenious invention! - <u>Consciousness, Information, Matter, Intelligence!</u> - This is the universal formula embedded in the system of the universe! - This is a system, with unchangeable law and principles! A system, which providing balance and stability of the universe, and dynamic recycling mechanism which ensures its continuation and eternity! In the system of the universe is embedded package of ethical principles - principles of non-interference, principles of self-determination, ethics, and freedom! It is a system with an unbreakable barrier for the advance of unethical societies! A system with principles preventing anybody from interfering with the balance and order of the universe and harming the others!

It will be easy to understand that the Universe and the term "God" are indivisible and we are an active part of this system, where the consciousness is the fundamental creative phenomena of our world.

We have to use our knowledge and understanding to be able to put harmony in our lives and get rid of corruption, arrogance, and social injustice. - This is the true aim of this book – to show what wrong we, the humans have been doing and believe in, and what we have to do to correct the tragic situation on our planet, where our technology can produce more than we can consume, but our ignorance, greed, and cruelty keep billions in poverty, hunger and misery and where we needlessly are destroying our planet and our future. It is time to realize that, if we want to call ourselves intelligent creatures, we have to stop doing these terrible things to ourselves and our planet. We have the technology to produce enough to make every single person happy without the necessity of destroying the planet or harm and rob each other!

The senseless greed of some individuals to accumulate billions and billions which they will never have a chance to spend in a lifetime is not a sign of intelligence, but exactly the opposite – primitive greed. It is time to realize that there is a super-intelligence behind the scenes which has created this world for us and giving us a chance to evolve into an intelligent, ethical, and moral society. We have to realize that this super-power has created order and rules of our universe which is not negotiable! The fact that we never observe even a single exception from the laws of physics and nature anywhere in the universe is actual proof that this world has been established under strict unchangeable principles! It is time for us to realize that we have to start obeying these principles! The time to become an intelligent and ethical society is overdue. The best description of our actions is that we are acting like vandals against ourselves and our planet. We have created a society based on lies and deception with no moral and ethical values. We are deceiving ourselves in every possible way with and without reason!

This book is different from the others because in simple and clear terms exposing the complexity of our total misunderstanding of the world, the wrong structure of the society and our wrong actions. This book is different because it does not simply criticize some subjects or actions, but gives a completely new vision for the structure of the world and our society - the real structure, without covering or twisting the truth. The book also provides the valuable knowledge - a new understanding of how to build our society, our life, and the correct understanding of who we are, where we come from, and what we have to do to make possible one more ethical, moral, prosperous and bright future for us our planet and for the next generations. - This knowledge can help and teach human civilization how to survive and prosper!

The earth is beautiful; it is our home! Don't lose it.

The author – Val Malinov is born in Bulgaria. His education includes engineering, electrical, radio communications, building and construction, home sustainability, music, art, and chemists.
He is the inventor of spherical combustion engine – 'Val Rotary Engine.'
Val has received a national bronze medal in whitewater slalom.
He escaped from the communist regime of his native country by a 3-day solo journey using a kayak across the Black Sea.
Val is a member of the Astronomical Society of Victoria.
His special interests are Physics, Quantum mechanics, astronomy, philosophy, classical music, art, and contact with nature.

The main context of the book is to shed light on the most vital questions of humanity: Where the Universe comes from? Is it spontaneous, or has it been created? What is the real origin of life? What is the real story behind the appearance of modern humans? Are we alone in the universe? Do we have a future?

To answer these questions, the book takes us on an unexpected journey through scientific considerations of the fundamentals of quantum mechanics, the clash of religion and science, the true facts of the Big Bang theory and considering never revealed facts of the most popular theories. The most unexpected revelation of the book is that the Universe is very logically constructed and we – the humans are an emerging intelligence that has an active role and special future in this structure, but there are conditions attached!

ISBN

www.ingramcontent.com/pod-product-compliance
Lightning Source LLC
Chambersburg PA
CBHW041157280326

41927CB00019BA/3381